Automotive Steering, Suspension, and Braking Systems

Automotive Steering, Suspension, and Braking Systems

Frank Thiessen   Davis Dales

Reston Publishing Company, Inc
A Prentice-Hall Company
Reston, Virginia

Library of Congress Cataloging in Publication Data

Thiessen, Frank   Dales, Davis
  Automotive brakes, suspension, steering, and alignment—
principles and service.

  Includes index.
1. Automobiles—Brakes—Maintenance and repair.
2. Automobiles—Springs and suspension—Maintenance
and repair.
3. Automobiles—Steering-gear—Maintenance and repair.
4. Automobiles—Wheels—Alignment.
I. Title.
TL269.T47   1982   629.2'4'0288   82-9084
ISBN 0-8359-0291-9   AACR2
ISBN 0-8359-0290-0 (pbk.)

©1983 by Reston Publishing Company, Inc.
A Prentice-Hall Company
Reston, Virginia 22090

10 9 8 7 6 5 4 3 2 1

Printed in the United States of America

Contents

Chapter 4 SUSPENSION SYSTEMS 127

Chapter 5 WHEELS AND TIRES 173

Chapter 6 STEERING SYSTEMS 201

Chapter 7 WHEEL ALIGNMENT AND BALANCE 241

Appendix CONVERSION CHARTS 263

INDEX 267

Preface

Proper service and repair are important to the reliable operation of all motor vehicles. The service procedures recommended and described in this text are effective methods of performing service operations. Some of these service operations require the use of tools specially designed for the purpose. The special tools should be used as recommended in the manufacturers' service manuals.

This text contains various general precautions, which should be read carefully in order to minimize the risk of personal injury or damage to the vehicle resulting from improper service methods. However, these general precautions are not exhaustive. The authors could not possibly know, evaluate, and advise the service trade of all conceivable ways in which service might be carried out or of the possible hazardous consequences of each method. Accordingly, anyone who uses any given service procedure or tool must first satisfy himself thoroughly that neither personal safety nor the safety of the vehicle will be jeopardized by the service method selected.

CAUTION

Automobiles contain many parts dimensioned in the metric system as well as in the customary system. Many fasteners are metric and are very close in dimension to familiar customary fasteners in the inch system. It is important to note that, during any maintenance procedures, replacement fasteners must have the same measurements and strength as those removed, whether metric or customary. (Numbers on the heads of metric bolts and on surfaces of metric nuts indicate their strength. Customary bolts use radial lines for this purpose, while most customary nuts do not have strength markings.) Mismatched or incorrect fasteners can result in vehicle damage or malfunction and possibly personal injury. Therefore, fasteners removed from the car should be saved for reuse in the same locations whenever possible. Where the fasteners are not satisfactory for reuse, care should be taken to select a replacement that matches the original. For information on any specific make or model, refer to the appropriate service manual.

The authors have made every effort to give proper credit to the sources of illustrations and other materials, and would be grateful for information from readers on errors or omissions in this regard.

Automotive Steering, Suspension, and Braking Systems

Chapter 1

Shop Routine

Figure 1-2. The running gear includes the wheels, suspension, steering, and braking systems as well as the frame.

PART 1 THE AUTOMOBILE

Development and improvement have resulted in the automobile and the industry undergoing constant change. Air pollution and energy conservation policies have required additional ongoing changes as dictated by government standards.

The automobile has been reduced in size and weight in order to consume less energy. Passenger safety has also received much attention, and many features on today's automobile reflect this concern.

All these factors have contributed to the need for more highly skilled and more knowledgeable automotive technicians, which has resulted in increased emphasis on standardized certification programs for automotive technicians. Anyone wishing to be an automotive technician must acquire as much skill and knowledge as possible in the automotive repair trade.

Figure 1-1. The automobile consists of two major components, the body and the chassis. The chassis includes the drive train and the running gear. This picture shows a typical four-door body assembly.

PART 2 SAFETY

Working in an automobile shop can be both interesting and rewarding. If you are the type of person who likes a variety of work, you will find it in shop work, where a large number of different service jobs and procedures are carried out. The variety of jobs and procedures, however, requires a high degree of awareness of the importance of safety. Safety is your job, everyone's job.

Safety in the shop includes avoiding injury to yourself and to others working near you. It also includes avoiding damage to vehicles in the shop and damage to shop equipment and parts. The following are some factors to consider in practicing shop safety.

Personal Safety Rules

1. Wear proper clothing. Loose clothing, ties, uncontrolled long hair, rings, etc., can be caught in rotating parts or equipment and cause injury. Wear the kind of shoes that provide protection for your feet; steel-capped work boots with nonskid soles are best. Keep clothing clean.

2. Use protective clothing and equipment where needed. Use rubber gloves and apron as well as a face mask when handling batteries. Protective goggles or safety glasses are used for grinder work and the like.

3. Keep hands and tools clean to avoid injury to hands and to avoid falling due to slipping when pulling on a wrench.

4. Do not use compressed air to clean your clothes. This can cause dirt particles to be embedded in your skin and cause infection. Do not point the compressed air hose at anyone.

5. Be careful when using compressed air to blow away dirt from parts. You should not use compressed air to blow dirt from brake parts since cancer-causing asbestos dust may be inhaled as a result.

6. Do not carry screwdrivers, punches, or other sharp objects in your pockets. You could injure yourself or damage the car on which you are working.

7. Never get involved in horseplay or other practical jokes. They can lead to injury.

8. Make sure that you use the proper tool for the job and use it the right way. The wrong tool or its incorrect use can damage the part on which you are working or cause injury.

9. Never work under a car or under anything else that is not properly supported. Use safety stands properly placed to work under a car and use a creeper.

10. Do not jack a car while someone is under it.

Figure 1-5. Face shields protect the face from possible injury resulting from acids, freon, flying particles from grinding wheels, chipping metal, and the like. *(Courtesy of Mac Tools Inc.)*

Figure 1-3. Protective leather gloves should be worn for welding. Rubber gloves protect hands when working with batteries. *(Courtesy of Mac Tools Inc.)*

Figure 1-6. Those requiring glasses to correct a visual defect should use glasses with safety lenses. *(Courtesy of Mac Tools Inc.)*

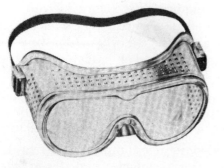

Figure 1-4. Safety goggles shield the eyes from injury. *(Courtesy of Mac Tools Inc.)*

Figure 1-7. Always support the car on properly placed jack stands before working under it. *(Courtesy of Walker)*

Figure 1-8. Use a creeper for working under a vehicle. *(Courtesy of Mac Tools Inc.)*

11. Never run a car engine without proper ventilation and adequate means of eliminating exhaust gases. Exhaust gas contains deadly carbon monoxide. It can and does kill.

12. Keep your work area clean at all times. Your safety and the quality of your work depend on it.

13. Lifting and carrying should be done properly to avoid injury. Heavy objects should be lifted and moved with the right equipment for the job.

14. Do not stand in the plane of rotating parts such as fans, etc. Always disconnect the electric fan before any underhood work. The fan may start any time underhood temperature rises above specified fan cut-in temperature, even with the engine off.

15. When working with others, note any unsafe practices and report them.

Shop Safety

1. Familiarize yourself with the way the shop is laid out. Learn where things are in the shop. You will need to know where the shop manuals are kept in order to obtain specifications and service procedures. Make sure that you know the route to the exit in case of fire.

2. Learn whether certain stalls are reserved for special jobs. Abide by these rules.

3. Take note of all the warning signs around the shop. No smoking signs, special instructions for some shop tools and equipment, danger zones, etc., are there to help make the shop run smoothly and safely.

4. Note the location of fire extinguishers. Take time to read their operating instructions and the type of fire on which they are meant to be used.

5. Follow local regulations when stor-

Figure 1-9. Dirty shop floors can cause major injury.

Figure 1-10. Proper lifting methods are a must to avoid back injury.

Figure 1-11. Avoid getting clothing caught in rotating parts such as fans, pulleys, grinders, and drills. Do not stand in the plane of rotating parts that are not shielded.

Figure 1-12. Electrical cords and connectors must be in good condition to avoid injury.

ing gasoline and other flammable liquids. Gasoline should be stored only in approved containers and locations.

6. Never use gasoline to clean parts. Never pour gasoline into a carburetor air horn to start the car.

7. Always immediately wipe up any gasoline that has been spilled.

8. Gasoline vapors are highly explosive. If vapors are present in the shop, have the doors open and the ventilating system on in order to eliminate these dangerous vapors.

9. Repair any gasoline leak immediately before doing any other work on a car. The potential fire hazard is very high. The least spark can set off a fire or explosion that is uncontrollable and fatal.

10. Dirty and oily rags should be stored in closed metal containers to avoid catching fire.

11. Keep the shop floor and work benches clean and tidy. Oil on the floor can cause serious personal injury.

12. Do not operate shop tools or equipment in unsafe condition. Electrical cords and connectors must be in good condition. Bench-grinding wheels and wire brushes should be replaced if defective. Floor jacks and hoists must be in safe, operating condition and should not be used above their rated capacity. The same applies to mechanical and hydraulic presses, drills, and drill presses. Draw the attention of your instructor or shop foreman to any unsafe equipment or conditions.

13. Extension cords should not pose a hazard by being strung across walkways.

14. Do not leave jack handles in the down position across the floor. Someone could trip over them.

15. Do not drive cars over electrical cords, which could cause short circuits.

First Aid

1. Make sure that you are aware of the location and contents of the first-aid kit in your shop.

2. Learn if there is a resident nurse in your shop or school, and learn where the nurse's office is.

3. If there are any specific first-aid rules in your school or shop, make sure that you are aware of them and follow them. You should be able to locate emergency telephone numbers—such as ambulance, doctor, and police—quickly.

4. There should be an eye-wash station in the shop to rinse your eye thoroughly should you get acid or some other irritant into it.

5. Burns should be cooled by rinsing with water immediately and then treated as recommended.

6. If someone is overcome by carbon monoxide, get the person to fresh air immediately.

7. In case of severe bleeding, try to stop blood loss by applying pressure with clean gauze on or around the wound and summon medical aid.

8. Do not move someone who may have broken bones unless life is further endangered. Moving a person may cause additional injury. Call for medical assistance.

PART 3 SHOP MANUALS, WORK ORDERS, AND PARTS LISTS

Shop Manuals

Shop manuals are a necessary part of the automotive service shop. They are needed to obtain the desired specifications and for specific service procedures. Mistakes and comebacks can be almost eliminated by the proper use of the correct shop manual.

Manufacturers' shop manuals are the most reliable source of information. Other available shop manuals often provide helpful hints and suggestions.

Figure 1-13 shows the index of three different sections in a manual. This directs the reader to the desired section, where there is another index (Figure 1-14). This index leads to the particular area for which information is being sought, for example, the manual and power steering gear specifications (Figure 1-15).

Work Orders

The sample work order illustrated in Figure 1-16 has room for the following information:

- Place of business
- Name of the customer
- Date of the work order
- Work order number
- Vehicle identification
- Type of service required
- Customer's signature
- List of parts used and their cost
- Labor costs
- Tax
- Responsibilities and liabilities of place of business and of the customer

The work order serves as a means of communication among the various parties involved in the repair procedure, such as the service writer, customer, technician, shop foreman, parts departments, cashier, and accounting department. The technician usually gets the hard copy, on which he records the time used to repair the vehicle and the parts required. The original copy is given to the customer on receipt of payment, and the remaining copies stay with the place of business for its records.

Parts Lists

Whether the hard copy of the work order is used for a parts list or a separate requisition is used, the parts department requires at least the following information in order to provide the correct parts for the unit being serviced: vehicle make, year, and model must be provided in every case.

Further information is included depending on which component of the vehicle is being serviced, for example,

- Suspension: standard or heavy-duty
- Steering: manual or power
- Brakes: standard or power, drum diameter, shoe width, wheel cylinder, or caliper piston diameter
- Rear axle: standard or traction type, ratio
- Engine: displacement, engine number, two- or four-barrel carburetor, single or dual exhaust, with or without air conditioning
- Carburetor: make, model, number of barrels

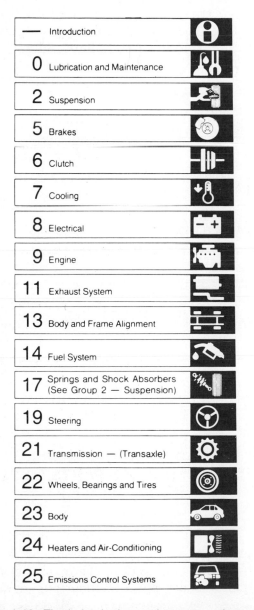

—	Introduction	
0	Lubrication and Maintenance	
2	Suspension	
5	Brakes	
6	Clutch	
7	Cooling	
8	Electrical	
9	Engine	
11	Exhaust System	
13	Body and Frame Alignment	
14	Fuel System	
17	Springs and Shock Absorbers (See Group 2 — Suspension)	
19	Steering	
21	Transmission — (Transaxle)	
22	Wheels, Bearings and Tires	
23	Body	
24	Heaters and Air-Conditioning	
25	Emissions Control Systems	

Figure 1-13. The index is the road map to a shop manual. It gets you where you want to go for information, specifications, and procedures. This is an example of a shop manual sectional index. It directs you to the appropriate section in the manual.

STEERING

CONTENTS

INDEX

Figure 1-14. Another index appears at the front of each section in a shop manual as shown here. *(Courtesy of Chrysler Corporation)*

SPECIFICATIONS

MANUAL STEERING GEAR

Fill gear with 0.11—0.14 litre (1/4 pint) of S.A.E. 90 hypoid oil
Apply silicone grease to tie rod bellows groove to prevent rotation of bellows.

TIGHTENING REFERENCE	Newton metres	Foot-Pounds
Clamp and Housing Pad Bolts	23-54	17-25
Tie Rod End Nut	34-70	25-50
Tie Rod End Lock Nut	60-90	45-65

POWER STEERING GEAR

Apply silicone grease to tie rod bellows groove to prevent rotation of bellows.

TIGHTENING REFERENCE	Newton metres	Foot-Pounds
Clamp and Housing Pad Bolts	23-54	17-25
Tie Rod End Nut	34-70	25-50
Tie Rod End Lock Nut	60-90	45-65
Inner Tie Rod	95	70

POWER STEERING PUMP

OUTPUT FLOW
88.3 to 114 ML/S (1.4 to 1.8 GPM)
at 1500 RPM and minimum pressure.

PRESSURE RELIEF
6.20 to 6.90 MPa
(900 to 1000 PSI)

Power Steering Oil Return Hose   LENGTH 254 mm (10 in.)

IF RETURN HOSE IS CHANGED, USE ONLY HYPALON MATERIAL, MOPAR PART NUMBER 3879925 OR EQUIVALENT.

POWER STEERING HOSES

	Newton metres	Foot-Pounds
Pressure Hose Tube Nuts (Both Ends)	20	15
Return Tube Nut	20	15
Pressure Hose Locating Bracket At Pump	40	30
Pressure Hose Locating Bracket At Crossmember	12	9
Return Tube Locating Bracket At Gear	28	21

POWER STEERING PUMP

	Newton metres	Foot-Pounds
Discharge Fitting	55	40
Relief Valve Ball Seat	6	4
Bracket Mounting Fasteners:		
3/8—16 Stud	48	35
3/8—16 Bolt and Nuts	40	30
M8 Bolts	28	21

Figure 1-15. Part of the specifications from a recent shop manual. *(Courtesy of Chrysler Corporation)*

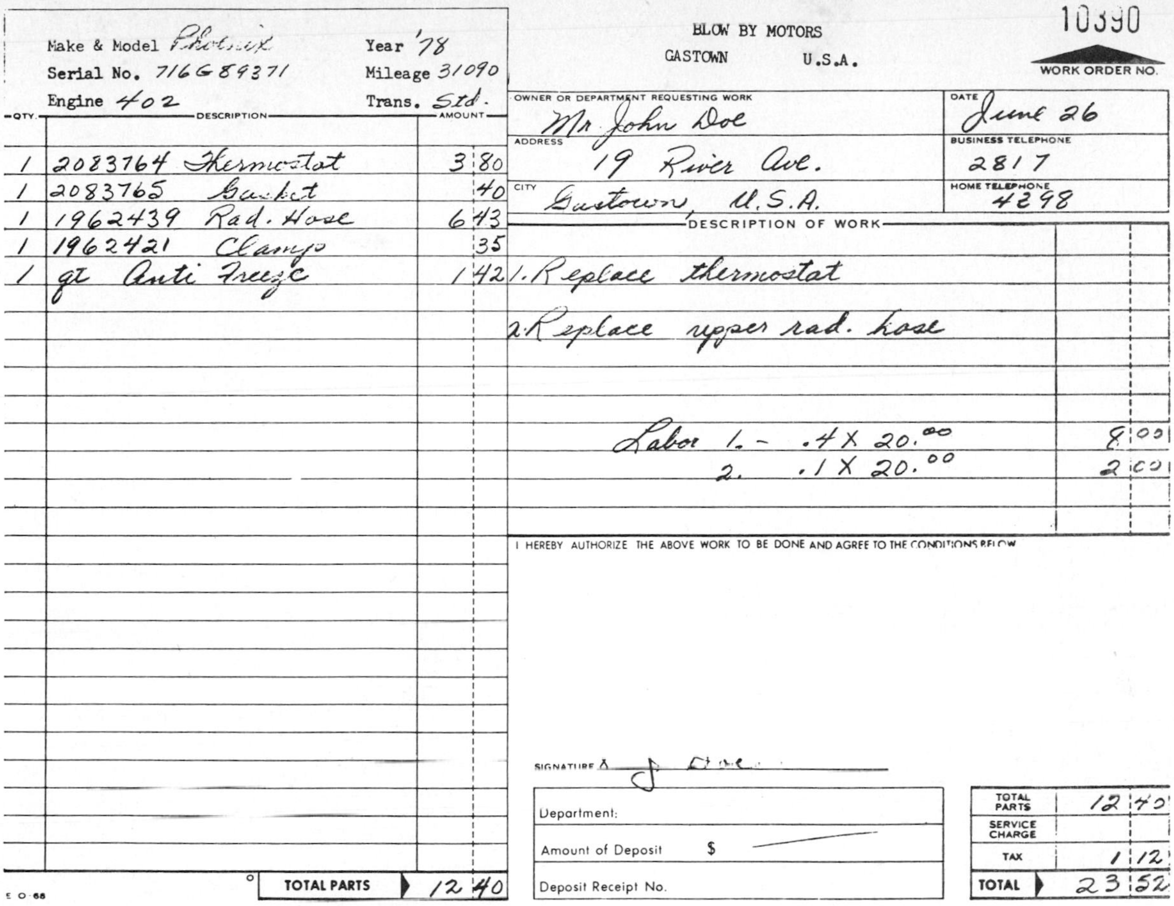

Figure 1-16. Sample copy of a work order or repair order.

• Transmission: type, standard or automatic transmission, number, and model

• Clutch: diameter, number of springs in pressure plate

Naturally, the correct names of the parts required must be used when ordering or requisitioning parts. When the correct name is unknown, it usually becomes quite difficult to communicate with the parts department. Learning the correct name for each part is one of the first things a prospective technician should do in order to function in the service industry.

PART 4 TOOLS AND SHOP EQUIPMENT

This part of the chapter covers the classification, identification, and proper use of tools and shop equipment common to most automotive service shops. Special tools are covered in those chapters dealing with the service procedures that require their use.

The technician's job is made easier by a good selection of quality tools and adequate shop equipment. The quality and speed of work are also increased. Fast and efficient work is necessary to satisfy the customer and the employer. An efficient, productive technician also experiences greater job satisfaction and earns more money as a result. The technician should not jeopardize ability by selecting tools that are inadequate or of poor quality. Good tools are easier to keep clean and last longer than tools of inferior quality.

Good tools deserve good care. Select a good roll cabinet and a good tool box to store your tools properly. They represent a fairly large investment and should be treated accordingly. Measuring instruments and other precision tools require extra care in handling and storage to prevent damage. Keeping your tools clean and orderly is time well spent. It increases your speed and efficiency on each job you do.

Hand Tools

Wrenches

Open-end, box-end, combination, and Allen wrenches are used to turn bolts, nuts,

7

Figure 1-17. Typical roll cabinet with tool box on top, providing good storage, easy access, and portability of tools. *(Courtesy of Mac Tools Inc.)*

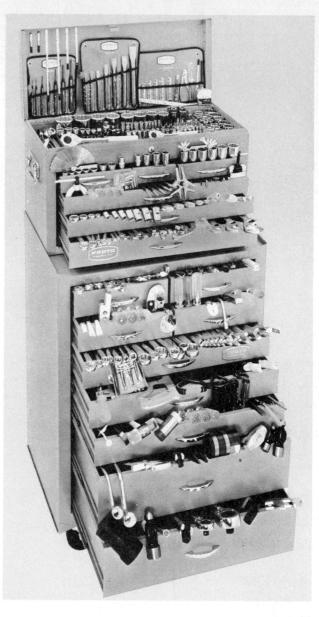

Figure 1-18. Example of a well-stocked roll cabinet and tool box for the professional technician. Can you name all the items shown? *(Courtesy of Proto Canada, Div. Ingersoll-Rand Canada Inc.)*

and screws. The open-end wrench holds the nut or bolt on only two flat sides. They slip or round off the nut more readily than do box-end wrenches. It is better to use box-end wrenches wherever possible. Various offsets are available to facilitate getting at tight places. Both 6- and 12-point box-end wrenches are available.

Sizes range from 3/8 to 1 1/4 inches in the average set, in 1/16-inch steps. Metric wrench sets range from 6 to 32 millimeters. Allen wrench sizes generally range from 2 to 20 millimeters and from 1/8 to 7/16 inch. Other sizes are also available.

Ratcheting box-end wrenches are very handy and come in similar size ranges as open-end and box-end wrenches. For tubing fittings, flare nut wrenches should be used rather than open-end wrenches. The tech-

nician should also have an adjustable wrench, but this wrench should not be used in place of the proper open-end or box-end wrench.

Figure 1-19. Heavy open-end wrench (above) and thinner open-end wrench (below). *(Courtesy of Mac Tools Inc.)*

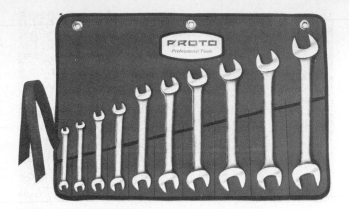

Figure 1-20. Open-end wrench set of various sizes in storage pouch. *(Courtesy of Proto Canada, Div. Ingersoll-Rand Canada Inc.)*

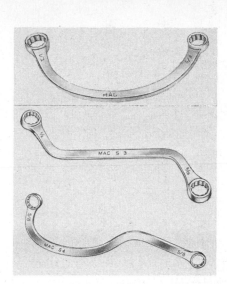

Figure 1-24. A number of special wrenches of various shapes for hard-to-get at places, such as manifolds, distributors, front end, and starters, are available and often essential. *(Courtesy of Mac Tools Inc.)*

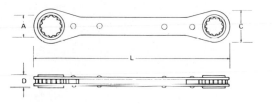

Figure 1-21. Box-end wrenches are available in many sizes in both 6- and 12-point styles. Various offset styles are also available. *(Courtesy of Mac Tools Inc.)*

Figure 1-22. Ratcheting box wrenches can be handy additions to any tool kit. *(Courtesy of Mac Tools Inc.)*

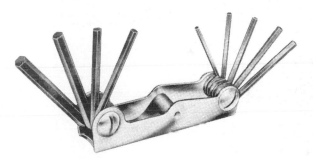

Figure 1-25. Hex or Allen wrenches are a must in the technician's tool kit.

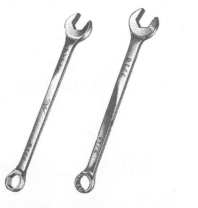

Figure 1-23. Combination wrenches have an open-end wrench in one end and a box-end on the other.

Figure 1-26. Flare nut or tubing wrenches should be used on tubing fittings to avoid rounding off the fittings.

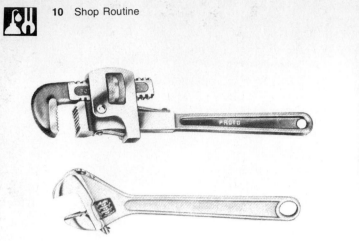

Figure 1-27. The adjustable wrench and the pipe wrench are necessary parts of a tool kit but should not be used in place of wrenches or sockets. *(Courtesy of Proto Canada, Div. Ingersoll-Rand Canada Inc.)*

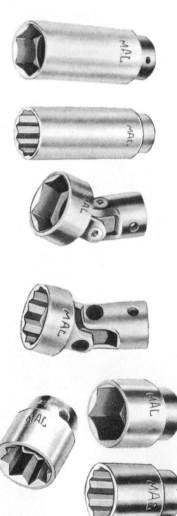

Figure 1-28. Standard, flex, and deep sockets are available in both 6- and 12-point types, as well as 8-point. *(Courtesy of Mac Tools Inc.)*

Sockets and Drives

The well-equipped automotive technician should have a 1/4-inch drive socket set, a 3/8-inch drive set, and a 1/2-inch drive set with standard and metric sockets. Socket wrenches are fast and convenient to use. Both 6- and 12-point sockets should be included in the well-equipped tool kit, as well as deep sockets and flex sockets. Socket sizes are similar to wrench sizes and metric sizes. Other drive sizes such as 3/4 or 1 inch are used for heavy-duty work.

Socket drives include universal joints, extensions of different lengths, ratchets, flex handles, T-handles, and speed handles. Drive sizes are available, as mentioned. Drive adapters are also available to increase or reduce drive sizes to fit available sockets.

A number of other socket attachments are available and are handy. Both electric and air-operated impact wrenches are used to drive sockets to speed the work; however, the sockets used are specially designed to withstand the continuous impacts of the driver.

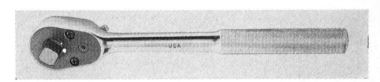

Figure 1-29. A ratchet allows turning the fastener without repeated removal of the wrench, thereby speeding up the work. Common drive sizes are 1/4, 3/8, 1/2 and 3/4 inch. *(Courtesy of Mac Tools Inc.)*

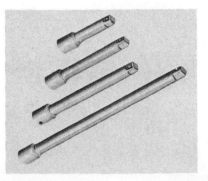

Figure 1-30. Drive extensions allow access to hard-to-get-at bolts and nuts. Extensions are available in various lengths. *(Courtesy of Mac Tools Inc.)*

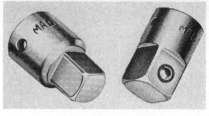

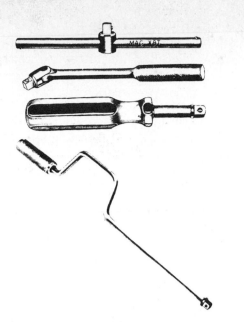

Figure 1-33. Drive size adapters convert socket drives up or down in size. Care must be exercised when using large driver on a small socket. Breakage can result. *(Courtesy of Mac Tools Inc.)*

Figure 1-31. Other socket drives include (from top to bottom) T-bar handle, flex handle, spinner, and speed handle.

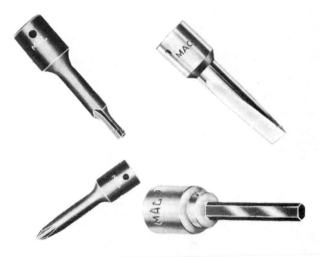

Figure 1-32. A universal joint converts the standard socket drive to a flex drive for hard-to-get-at places.

Figure 1-34. Other socket attachments include a standard screwdriver tip, a Phillips screwdriver, clutch screwdriver, and hex wrench. *(Courtesy of Mac Tools Inc.)*

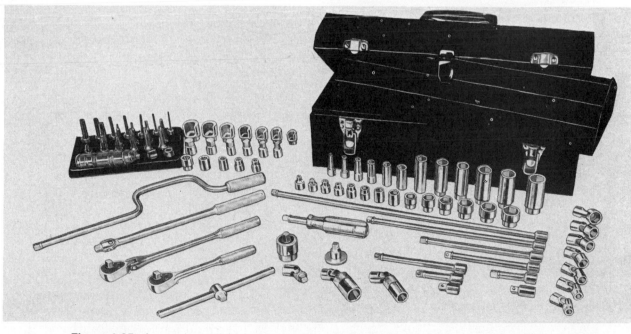

Figure 1-35. A complete 3/8-inch drive socket set including a swivel head ratchet and other special attachments.

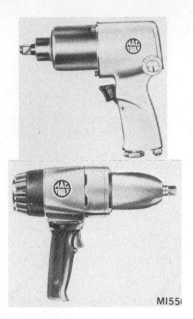

Figure 1-36. Impact tools speed the technician's work. Both air-operated (top) and electric impact tools (bottom) are available. *(Courtesy of Mac Tools Inc.)*

Figure 1-37. Special sockets such as these should be used with impact tools. They are able to withstand the heavier use and continuous impacts. *(Courtesy of Mac Tools Inc.)*

Torque Wrenches

To tighten a bolt or nut to specifications, it is necessary to use a torque wrench. Overtightening a bolt or nut puts excessive strain on the parts and damages the bolt or nut as well. On the other hand, if not tightned enough, the unit may come apart during use. Torque can be defined as twisting force. A 1-pound pull on a 1-foot wrench (from the center of the bolt to the point of pull) equals 1 pound-foot. Force times distance equals torque (F x D = T) or 1 pound x 1 foot = 1 pount-foot of torque.

Other values for force are used, such as kilograms (kiloponds) or newtons. In this case, the values used for distance are centimeter and meter. The amount of torque, depending on the type of torque wrench used, is measured in pounds-inches or pounds-feet for some cars. For imports, the centimeter-kilogram and meter-kilogram torque values are sometimes used. Another term for kilogram, kilopond, is used in connection with torque values. The current International System of Units (SI) torque values are given in newton-meters. This system is also being used by domestic manufacturers.

Use the appropriate type of torque wrench for the specifications given. If specifications are given in centimeter-kilograms, then a torque wrench with a centimeter-kilogram scale should be used. If torque values stated do not match the values on your torque wrench, use the metric conversion chart to convert the values given (see Appendix). Abbreviations for the various torque values given are as follows:

Pound-inches: lb-in.

Pound-feet: lb-ft

Centimeter kilograms: cm kg or cm kp

Meter kilograms: m kg or m kp

Newton meters: N • m

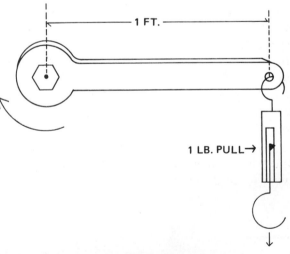

Figure 1-38. Torque is twisting force. Amount of torque is calculated by multiplying force times distance (F x D = T).

Figure 1-39. The dial-type torque wrench is available in various drive sizes. Dials are read directly and must be closely observed to torque fasteners correctly.

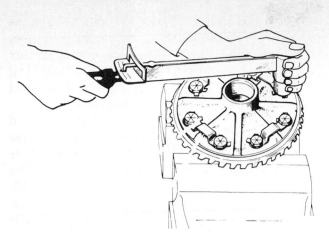

Figure 1-40. The scale-type torque wrench is read directly just as the dial type. A feelable-audible click attachment is provided on some models that signals when predetermined-preset torque has been reached.

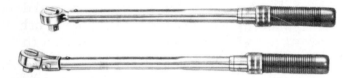

Figure 1-41. Another click-type torque wrench has a micrometer-type adjustment. There is no direct reading scale. Desired torque must be set on the micrometer adjustment. Preset torque is reached when the wrench clicks. *(Courtesy of Mac Tools Inc.)*

Screwdrivers

Screwdrivers are probably abused more than any other tool. Use them only for the purpose for which they were intended. There is no all-purpose screwdriver. Use the right type and size of screwdriver for the job. Slotted screws require flat-blade screwdrivers. Select a screwdriver with a tip that is as wide as the screw slot is long.

Use the correct size of Phillips, Reed and Prince, Robertson, or clutch-type screwdriver. Never make do with the wrong size.

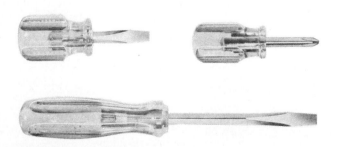

Figure 1-42. Common screwdriver types. The square socket type is also known as the Robertson screwdriver.

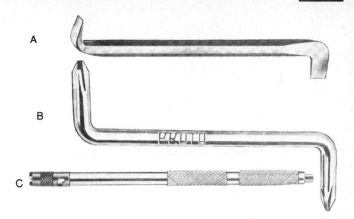

Figure 1-43. (A) Offset tip. (B) Phillips screwdrivers. (C) Screw-holding screwdriver for starting screws. *(Courtesy of Pronto Canada, Div. Ingersoll-Rand Canada, Inc.)*

Pliers

There are two groups of pliers. One is used for gripping and the other for cutting. Diagonal cutting pliers are sometimes called side-cutting pliers or side cutters. Gripping pliers should not be used in place of wrenches or sockets since this damages nuts and bolts. Do not grip machined or hardened surface parts with pliers; it will damage the surface.

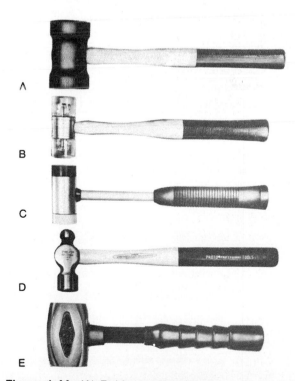

Figure 1-44. (A) Rubber mallet. B) Plastic Mallet hammer. (C) Rubber- or plastic-faced hammer. (D) Ball-peen hammer. (E) Sledge hammer. *(Courtesy of Pronto Canada, Div. Ingersoll-Rand Canada, Inc.)*

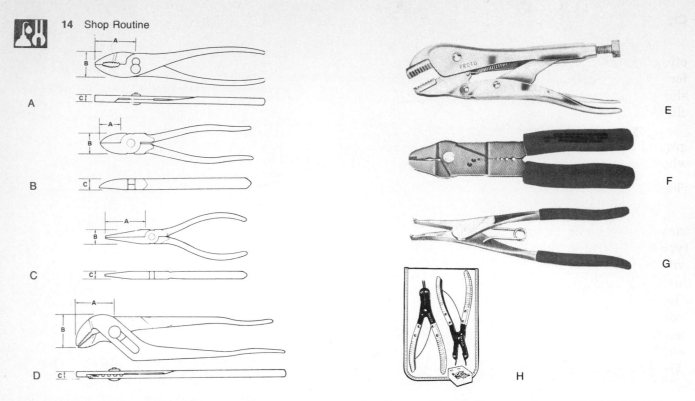

Figure 1-45. (A) Combination pliers. (B) Diagonal or side cutting pliers. (C) Needle-nose pliers. (D) Channel lock pliers. (E) Lock grip pliers. (F) Wire cutting and stripping tool. (G) External snap ring pliers. (H) Lock ring plier set with interchangeable tips.

Hammers

Automotive technicians use ball-peen hammers and soft hammers. Soft hammers such as plastic, rawhide, lead, or brass types are used on easily damaged surfaces.

A hammer should be held at the end of the handle. The hammer should land flat on the surface being struck.

Handles should be kept secure in the hammer head to avoid injury and damage. Select the right size (weight) of hammer for the job.

Punches

Pins and rivets are removed with punches. A tapered starting punch is used to start rivet removal after the rivet head has been chiseled or ground off. The rivet is then driven out the rest of the way with the pin punch. A long, tapered punch is used for aligning parts. A center punch is used to mark parts before disassembly and to mark the spot where a hole is to be drilled.

Punches should be kept in good condition. Do not allow mushrooming to take place.

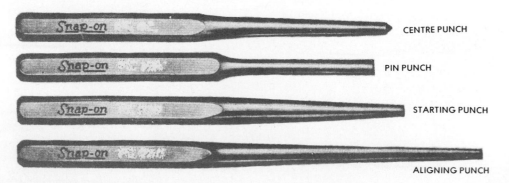

Figure 1-46. Common types of punches used by automotive technicians. *(Courtesy of Snap-on Tools Corporation)*

Chisels

Chisels are used to cut rivet heads and other metal. A chisel holder can be used for heavy work. Chisels should be kept sharp. Sharpen at approximately a 60° included angle.

Any sign of mushrooming should be ground off. Use safety goggles or a face mask when using a chisel.

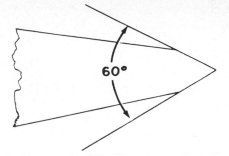
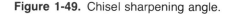

Figure 1-49. Chisel sharpening angle.

Files

Files are cutting tools used to remove metal, to smooth metal surfaces, etc. Many types of files are available for different jobs, with different sizes, shapes, and coarse or fine cutting edges. Size is determined as shown in Figure 1-51. The range of coarseness in order from coarse to fine is rough, coarse, bastard, second cut, and dead smooth. Never use a file without a handle. Do not use a file as a pry. Files are brittle and break easily.

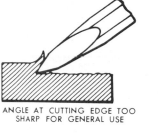
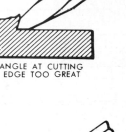

ANGLE AT CUTTING EDGE TOO SHARP FOR GENERAL USE

ANGLE AT CUTTING EDGE TOO GREAT

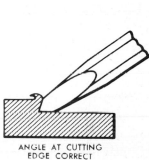

EDGE DULLED AND ROUNDED

ANGLE AT CUTTING EDGE CORRECT

Figure 1-47. Punch or chisel holder avoids injury to hands in heavy work.

Figure 1-50. Incorrect and correctly sharpened chisel angles. *(Courtesy of Ford Motor Co. of Canada Ltd.)*

Figure 1-48. Common chisel types that should be included in the technician's tool kit. *(Courtesy of Snap-on Tools Corporation)*

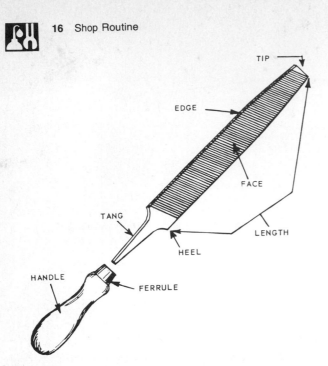

Figure 1-51. Parts of a file. Size is determined by length.

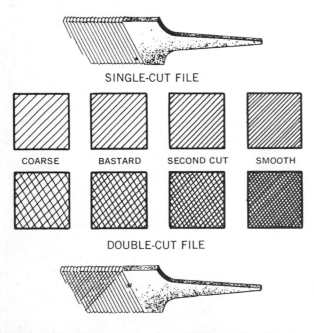

SINGLE-CUT FILE

COARSE   BASTARD   SECOND CUT   SMOOTH

DOUBLE-CUT FILE

Figure 1-52. Coarse to fine files, left to right. Single-cut (upper) and double-cut (lower) file types.

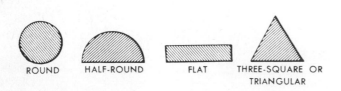

ROUND   HALF-ROUND   FLAT   THREE-SQUARE OR TRIANGULAR

Figure 1-53. Cross-sectional shapes of files. *(Courtesy of Ford Motor Co. of Canada Ltd.)*

Hacksaws

A hacksaw consists of an adjustable frame with a handle and a replaceable hacksaw blade. Blades are commonly available in 10- and 12-inch lengths. The coarseness of the cutting teeth is stated in number of teeth per inch, usually 14, 18, 24, or 32 teeth per inch. The finer blades are used for materials of thin cross section. Select a blade that will have at least 2 teeth cutting at all times.

Apply light pressure on the forward stroke and release the pressure on the return stroke. When replacing a hacksaw blade, install it with the teeth pointing away from the handle.

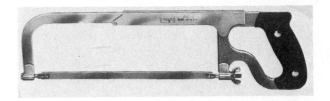

Figure 1-54. Adjustable frame hacksaw.

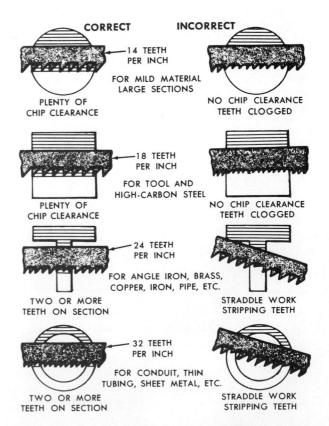

Figure 1-55. Correct and incorrect use of different hacksaw blades. *(Courtesy of Ford Motor Co. of Canada Ltd.)*

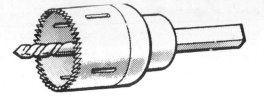

Figure 1-56. Power-driven hole saw. *(Courtesy of Ford Motor Co. of Canada Ltd.)*

Figure 1-59. Power-driven rotary wire brushes. *(Courtesy of Proto Canada, Div. Ingersoll-Rand Canada Inc.)*

Cleaning Tools

Hand-held scrapers are used to scrape the gasket or other material from flat surfaces. This should be followed by a light sanding.

The wire brush is used to clean rough surfaces. A soft bristle brush is used to help clean parts being washed in solvent.

Tubing Tools

Steel and copper tubing should be cut with a tubing cutter. Avoid applying too much pressure during cutting. This can collapse the tube. Low-pressure lines need only a single flare, whereas brake lines require double-lap flaring or International Standards Organization (ISO) flaring. Some vehicles have the double-lap flared brake lines while others have the ISO-type flared lines and fittings. Fittings and lines of different designs should never be mixed, used together, or interchanged.

After cutting, ream the tubing and then flare as required. The reamer is usually part of the tubing cutter. Make sure that no metal chips are left in the tubing.

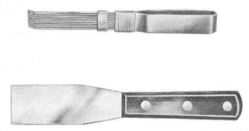

Figure 1-57. Parts cleaning scrapers: (A) for irregular surfaces; (B) for flat surfaces. *(Courtesy of Proto Canada, Div. Ingersoll-Rand Canada, Inc.)*

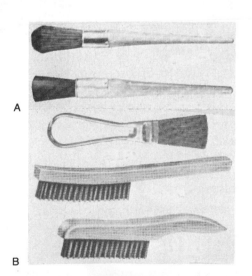

Figure 1-58. (A) Brushes for cleaning parts in solvent. (B) Wire brushes. *(Courtesy of Proto Canada, Div. Ingersoll-Rand Canada Inc.)*

Figure 1-60. Flaring tool kit required to flare tubular lines. *(Courtesy of Proto Canada, Div. Ingersoll-Rand Canada Inc.)*

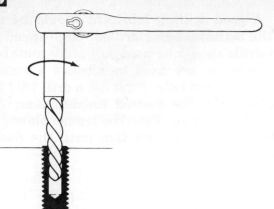

Figure 1-71. Using a screw extractor to remove a broken stud after it has been properly drilled.

Bushing and Seal Tools

There are many different types of bushing and seal drivers and pullers. Some automotive jobs require very special types of pullers or drivers. Pullers are either of the threaded type or the slide hammer type. Care must be exercised in selecting the correct type for the job.

Bushings and seals are very easily damaged if an improper tool or method is used. Refer to the toolmaker's instructions and the shop manual for specific information.

Miscellaneous Tools

A great variety of miscellaneous tools is available for the automotive service industry. Many of these can lighten and speed the task of the technician.

A magnetic retrieving tool is very handy. A stethoscope helps to localize and identify abnormal knocks and other noises on a vehicle. Essential miscellaneous tools include extension lights and cords, a good creeper, and fender covers.

A good technician will be aware of any new developments in tools and equipment that will improve the quality and productivity of work.

Soldering Irons

The technician may be required to solder such items as radiators and wiring connections. To do a good job of soldering, the surfaces must be smooth and absolutely clean. A good soldering iron of the proper size should be

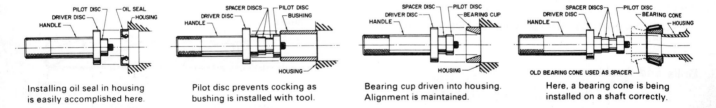

Installing oil seal in housing is easily accomplished here.

Pilot disc prevents cocking as bushing is installed with tool.

Bearing cup driven into housing. Alignment is maintained.

Here, a bearing cone is being installed on a shaft correctly.

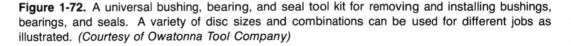

Figure 1-72. A universal bushing, bearing, and seal tool kit for removing and installing bushings, bearings, and seals. A variety of disc sizes and combinations can be used for different jobs as illustrated. *(Courtesy of Owatonna Tool Company)*

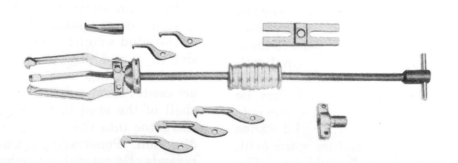

Figure 1-73. A slide hammer puller kit with attachments. Inside and outside jaws are interchangeable. *(Courtesy of Owatonna Tool Company)*

used: heavier irons for heavier work, a soldering gun for electrical work, etc. The soldering iron should be thoroughly cleaned and tinned (coated with solder). Solder is available in bar or wire form. The correct flux must also be used. Rosin flux must be used for electrical work to prevent corrosion. Acid flux is used on other work. Acid-core solder and rosin-core wire solder are most frequently used.

Soldering irons should be placed in holders during heating and cooling.

Figure 1-77. Electric soldering iron. *(Courtesy of Snap-on Tools Corporation)*

Figure 1-74. A long-handle, flex-head magnet for retrieving parts.

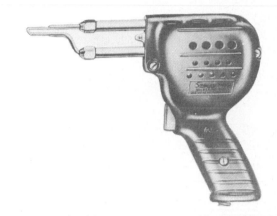

Figure 1-78. Electric soldering gun used for soldering electrical connections. *(Courtesy of Snap-on Tools Corporation)*

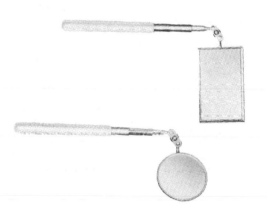

Figure 1-75. Mirrors with flex heads and long handles allow the technician to look at otherwise inaccessible places.

Figure 1-79. Bar- and wire-type solder.

Measuring Tools

All measuring tools should be treated as precision instruments. They should be properly used, cared for, and stored. Tools provided with special storage cases should be put in the case when not in use. Be careful not to drop precision tools. Do not leave them lying around, among or under other tools. Precision surfaces, such as straightedges, micrometer anvils, and caliper

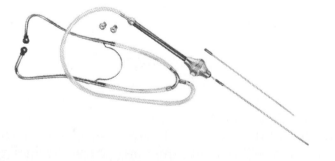

Figure 1-76. The stethoscope is used to determine the exact location of knocks and other abnormal noises.

jaws, should not be marred, scratched, or dented. The accuracy of measuring instruments should be checked if any doubt exists. They can be checked against other tools known to be accurate. Gauge blocks for micrometers determine micrometer accuracy. If no means of checking are available in the shop, they should be sent out to be checked.

Figure 1-80. Standard English feeler gauge with 10 blades from 0.002 to 0.015 inch. *(Courtesy of Mac Tools Inc.)*

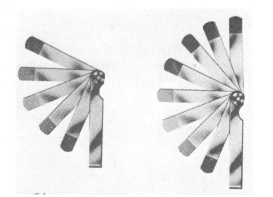

Figure 1-81. "Go-no go" gauge with stepped feeler blades.

Feeler Gauges

Feeler gauges are precision measuring tools used to measure small clearances. Various blade lengths are available. Flat feeler gauges have their thickness marked in thousandths of an inch, millimeters, or both. Nonmagnetic feeler gauges are used for checking clearances where magnetic force exists. Stepped feeler gauges have the tip of the blade 2/1,000 inch thinner than the rest of the blade. These are used for "go-no go" quick measurements.

Never bend, twist, force, or wedge feeler gauges since this destroys their accuracy. Wipe blades clean with an oily cloth to prevent rust.

Calipers

Inside and outside calipers are useful in

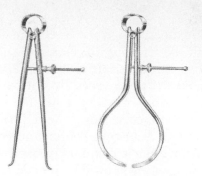

Figure 1-82. Inside and outside calipers. *(Courtesy of The L.S. Starrett Company)*

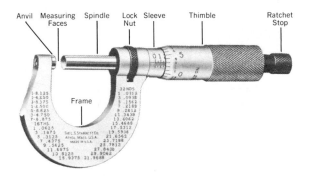

Figure 1-83. Outside micrometer with major parts identified. A piece of work is measured by placing it between the anvil and the spindle faces and turning the spindle by means of the thimble until both faces contact the work. The micrometer reading is then taken. *(Courtesy of The L.S. Starrett Company)*

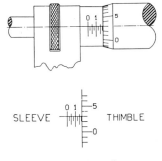

READING .178"

Figure 1-84. Micrometer reading (English scale).

measuring to an accuracy of approximately 1/100 inch. Vernier calipers measure inside and outside dimensions. English, metric, and combination types are available. English vernier calipers measure to an accuracy of 0.001 inch, while metric calipers measure to an accuracy of 0.02 millimeters. Most vernier calipers also have a depth-measuring rod attached.

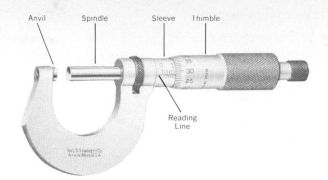

Anvil Spindle Sleeve Thimble

Reading
Line

Figure 1-85. The metric micrometer is similar in appearance to the English micrometer. *(Courtesy of The L.S. Starrett Company).*

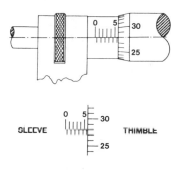

SLEEVE   THIMBLE

Reading 5.78 mm

Figure 1-86. Metric micrometer scale.

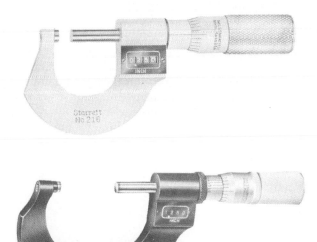

Starrett
No. 216

Starrett
No. 216

Figure 1-87. Digital readout micrometers are quicker to read but are more expensive.

Micrometers

Outside micrometers are used to measure the size of parts, such as diameter and thickness. The size of a micrometer is determined by the distance between the face of

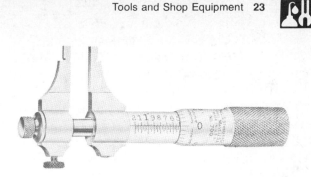

Figure 1-88. Inside micrometers in English and metric are read the same as outside micrometers. *(Courtesy of The L.S. Starrett Company)*

the anvil and the face of the spindle when the micrometer is adjusted to its minimum and maximum adjustments. A 1- to 2-inch micrometer would not be able to measure anything less than 1 inch or anything over 2 inches. A 25- to 50-millimeter micrometer would not be able to measure anything less than 25 millimeters or anything over 50 millimeters.

Digital micrometers measure inside and outside diameters, are available in sizes and types similar to conventional micrometers, and are read directly.

English Micrometer Scale.

Each division on the sleeve represents 0.025 or 25/1,000 inch. From 0 to 1 represents four such divisions (4 x 0.025 inch) or 0.100 inch (100/1,000 inch). Each division on the thimble represents 0.001 or 1/1,000 inch. There are 25 such divisions on the thimble. Therefore, one turn of the thimble moves the spindle 0.025 or 25/1,000 inch. To read the micrometer in thousandths, multiply the number of divisions visible on the sleeve by 0.025, then add the number of thousandths indicated by the line on the thimble, which coincides with the long reading line on the sleeve.

Example: Refer to Figure 1-84.

The 1 line on sleeve is visible, representing   0.100

There are three additional lines visible, each representing 0.025 inch.   0.075

Line 3 on the thimble coincides with the reading line on the sleeve, each line representing 0.001 inch.   <u>0.003</u>

The micrometer reading is   0.178 inch

Metric Micrometer Scale.

One revolution of the thimble advances the spindle toward or away from the anvil a 0.5 millimeter distance.

The reading line on the sleeve is graduated in millimeters (1.0 millimeter) from 0 to 25. Each millimeter is also divided in half (0.5 millimeter). It requires two revolutions of the thimble to advance the spindle 1.0 millimeter. The beveled edge of the thimble is graduated in 50 divisions from 0 to 50. One revolution of the thimble moves the spindle 0.5 millimeter. Each thimble graduation equals 1/50 of 0.5 millimeter or 0.01 millimeter.

To read the micrometer, add the number of millimeters and half-millimeters visible on the sleeve to the number of hundredths of a millimeter indicated by the thimble graduation that coincides with the reading line on the sleeve.

Example: Refer to Figure 1-86.

The 5-millimeter sleeve graduation is visible.	5.00
One additional 0.5-millimeter line is visible on the sleeve.	0.50
Line 28 on the thimble coincides with the reading line on the sleeve; 28 x 0.01.	0.28
The micrometer reading is	5.78 mm

Telescoping Gauges

Telescoping gauges can be used in conjunction with outside micrometers for measuring inside diameters, instead of using inside micrometers.

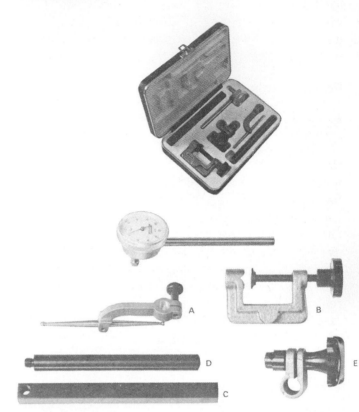

Figure 1-90. This dial indicator set takes care of many measuring requirements in the shop. Dial indicators are used to measure differential gear backlash, disc and flywheel runout, crankshaft and transmission shaft end play, and so on. (A) hole attachment or wiggle bar; (B) clamp; (C) toolpost holder; (D) upright spindle; (E) universal clamp. *(Courtesy of The L.S. Starrett Company)*

Range, English	Range, Metric
5/16 to 1/2 in.	7.9 to 12.7 mm
1/2 to 3/4 in.	12.7 to 19 mm
3/4 to 1-1/4 in.	19 to 31.7 mm
1-1/4 to 2-1/8 in.	31.7 to 54 mm
2-1/8 to 3-1/2 in.	54 to 89 mm
3-1/2 to 6 in.	89 to 152.4 mm

Figure 1-89. Telescoping gauges are used to measure inside diameters when used in conjunction with an outside micrometer. The knurled handle releases the spring-loaded plungers to diameter size, then they are locked by turning the knurled handle. The telescoping gauge is then carefully removed and measured with an outside micrometer. *(Courtesy of The L.S. Starrett Company)*

Dial Indicators

The dial indicator is needed to perform many of the measurements required in the shop. Graduations on the scale are in either 0.001 inch (1/1,000 inch) or in 0.01 millimeter (1/100 millimeter) for general shop use. The measurement range of the dial indicator is determined by the amount of plunger travel provided.

Power Tools

Bench Grinder

The bench grinder is an indispensable item in any shop. Usually one end of the motor shaft has a grinding wheel and the other has a wire wheel. The grinder can then be used for sharpening tools and cleaning parts. Other jobs for the grinder are grinding rivets or removing stock from metal parts. Grinding wheels of different types and sizes are available for specific jobs. Grinding and cleaning require skill and careful handling to avoid injury to the operator or the tools and parts being reworked.

Electric Drills

Hand-held electric drills perform a variety of jobs, as listed in Figure 1-93. When using a hand-held drill, a few general rules will make the job easier. Do not apply any side pressure when drilling since this can break drill bits. Apply only enough pressure for good drilling; too much pressure can cause overheating and destroy the drill bit. Ease up on the pressure just before the drill breaks through to prevent grabbing. Keep a firm grasp on the drill at all times. Use a small amount of cutting oil when drilling steel. Make sure that the piece of work being drilled is held securely.

These general rules apply to the drill press as well. The drill press has a movable table that can be raised, lowered, and turned sideways. A drilling block should be used on top of the table to avoid drilling into the table.

Figure 1-93. Hand-held electric drill. Common types include 1/4, 3/8, and 1/2 inch. Available in low-speed, high-speed, and variable-speed models. Common uses are for drilling, honing, and driving cleaning brushes, hole-saws, and rotary files. *(Courtesy of Mac Tools Inc.)*

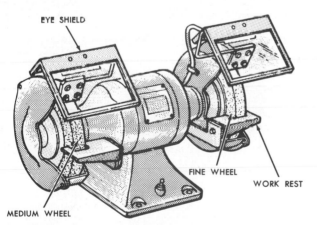

Figure 1-91. Bench grinder with tool supports and protective shields in place. *(Courtesy of Ford Motor Co. of Canada Ltd.)*

Figure 1-92. Wire wheel and grinding wheel attachments for the bench grinder.

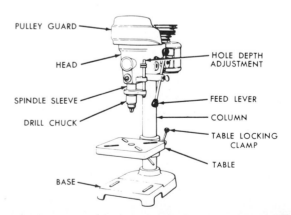

Figure 1-94. The drill press provides for a number of drilling speeds by changing the drive belt to different pulley positions. The drill press is used for precision-drilling of parts that can be carried by hand. *(Courtesy of Ford Motor Co. of Canada Ltd.)*

Jacks, Hoists, Lifts, and Presses

Hydraulic Jacks

Hydraulic floor jacks, transmission jacks, and bumper jacks are used in the automotive shop. The floor jack is used to raise a vehicle at the front, rear, or sides. The jack should be properly placed so that under-vehicle parts are not damaged. The floor jack can also be used to help move cars in tight places.

The bumper jack should only be used on bumpers that can withstand the strain. Place pads at bumper brackets. Do not work under a car supported only on a jack. Always use jack stands properly placed.

Figure 1-97. Bumper jack can be used to raise vehicles. Pads must be properly positioned and bumper must be of adequate strength; commonly used when changing wheels and tires in the shop.

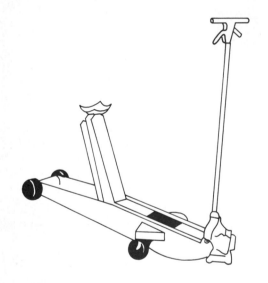

Figure 1-95. Portable hydraulic floor jack used to raise vehicles.

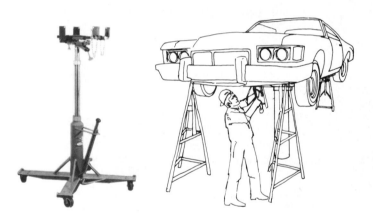

Figure 1-98. Under-hoist stands used to spread springs, unload ball joints, facilitate exhaust system replacement, and be a safety device.

Hoists

Different types of hoists are used to raise vehicles for under-vehicle work. A good hoist allows easy vehicle positioning and has minimal under-vehicle obstruction. Single-post, frame-contact hoists and twin-post and drive-on hoists are commonly used in automotive shops.

Chassis Lubrication Equipment

Chassis lubrication equipment is located in the lube bay and provides a motor oil dis-

Figure 1-96. Always support raised vehicle with properly placed jack stands before doing any work under vehicle.

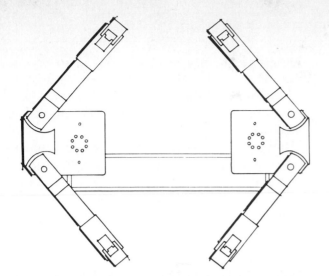

Figure 1-99. Single-post frame contact hoist. Arms and pads pivot for positioning under proper lift points of vehicle.

Figure 1-100. One type of drive-on hoist with frame lift attachment extended.

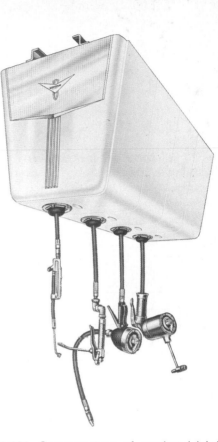

Figure 1-101. Common type of overhead lubrication dispensing equipment. This unit has an air hose, grease gun, gear oil dispenser, and motor oil dispenser. This type of equipment is located in the lube bay, which is also equipped with a hoist. *(Courtesy of Lincoln, St. Louis)*

penser, gear oil dispenser, grease gun, and air pressure hose. Different car manufacturers recommend lubrication at different mileage intervals. Lubrication points generally include tie rod ends, ball joints, and some suspension components. The number of points requiring lubrication varies among different vehicle makes. Some makes have lubrication fittings already installed; others have plugs that must be removed and lube fittings installed before lubrication can be done.

Follow the manufacturer's recommendations and specifications for periodic lubrication and types of lubricants required.

Hydraulic Presses

The hydraulic press should not be used beyond its rated capacity. All work should be properly positioned and supported, and all recommended shields and protective equipment should be used. The hydraulic press exerts tremendous pressure and can cause parts under pressure literally to explode, causing serious injury.

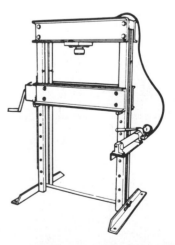

Figure 1-102. Floor-type and bench-top hydraulic shop presses. Floor model has an adjustable press bed for different working positions. Presses are used to remove and install bearings, gears, bushing, piston pins, and so on.

Figure 1-103. The bench vise is used to hold items on which work is being done. The soft jaw at right must be used when clamping easily damaged surfaces. *(Courtesy of The L.S. Starrett Company)*

PART 5   FASTENERS

A great variety of types and sizes of fasteners is used in the automotive industry. Each fastener is designed for a specific purpose and for specific conditions encountered in vehicle operation.

Using an incorrect fastener or a fastener of inferior quality for the job can result in early failure and even injury to the driver and passengers.

Some precautions to observe when replacing fasteners are the following.

• Always use the same diameter, length, and type of fasteners as originally used by the vehicle manufacturer.

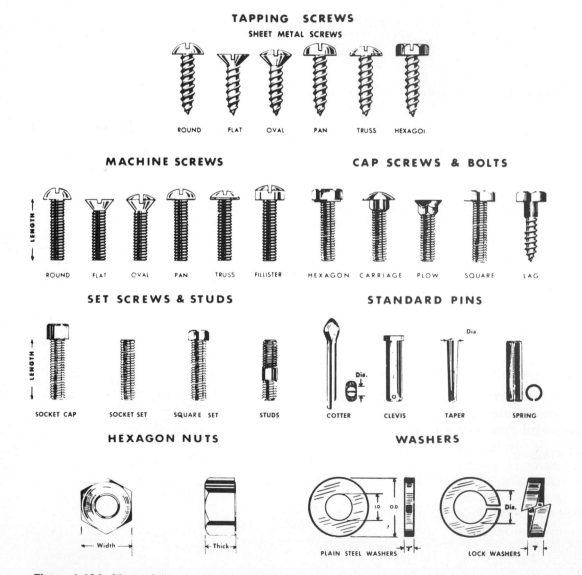

Figure 1-104. Many of the common types of fasteners used in the automotive industry. *(Courtesy of H. Paulin & Co. Limited)*

• Never thread a fastener of one thread type to a fastener of a different thread type.

• Always use the same number of fasteners as used originally by the manufacturer of the vehicle.

• Always observe the vehicle manufacturer's recommendations for tightening sequence, tightening steps (increments), and torque values.

• Always use the correct washers, pins, and locks as specified by the vehicle manufacturer.

• Always replace stretched fasteners or fasteners with damaged threads.

• Never use a cotter pin more than once.

Damaged threads in threaded parts can be restored by the use of helically coiled thread inserts. Replace damaged snap rings and keys with new ones. The completed work is only as good as the technician's desire and ability to do a professional job with the use of correct parts and fasteners.

A number of terms have been used over the years to identify the various types of threads. Some of these have been replaced with new terms. The terms most commonly used in the automotive trade follow.

The United States Standard (USS), the American National Standard (ANS), and the Society of Automotive Engineers Standard (SAE) all have been replaced by the Unified National Series. The Unified National Series consists of four basic classifications:

1. Unified National coarse (UNC or NC)

2. Unified National fine (UNF or NF)

3. Unified National extrafine (UNEF or NEF)

4. Unified National pipe thread (UNPT or NPT)

The two common metric classifications are coarse and fine and can be identified by the letters SI (Systeme International d'Unites or International System of Units) or ISO (International Standards Organization).

Wheel Studs and Nuts

Automobile wheels are usually mounted with four or five specially designed studs and nuts. They are a tight fit in the wheel hub or axle flange. This part of the stud is usually serrated to provide the necessary interference fit and to prevent the stud from turning when the wheel nuts are tightened. Nuts are usually coned on the wheel side of the nut to fit corresponding tapered holes in the wheels. This helps to position and center the wheel on the hub and helps to keep them tight by the wedging action of this design.

Wheel nuts for magnesium or aluminum wheels are usually provided with special washers to prevent fracturing the wheel metal from the friction of the nut during tightening. Some wheel designs use wheel mounting bolts or cap screws with tapered heads and threaded wheel hubs.

Most wheel studs and nuts use right-hand threads but some designs may use left-hand threads on the left side of the vehicle. Many vehicles use the customary thread but others use ISO metric threads. Wheel studs and nuts are critical safety attaching devices, which should be treated with care. Different thread types or designs should never be mixed; all wheel nuts should always be tightened in the specified sequence and to the torque specified by the vehicle manufacturer.

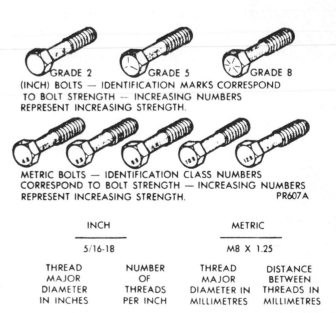

Figure 1-105. SAE and metric grade and thread identification. SAE bolt head markings show 2 lines less than its SAE grade; that is, bolt with 6 lines is SAE grade 8. Common metric grades are 4.8, 8.8, 9.8, 10.8, 12.9, as shown. An SAE grade 5 is the equivalent of a metric grade 8.8.

HEXAGON   HEXAGON WASHER FACED   SQUARE (CHAMFER)   PLAIN

SELF LOCKING NUT   WHEEL NUT   WING NUT   HEX. CAP.

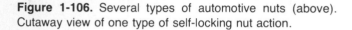

LOCKING ACTION:

Mating Threads of Bolt and Nut Wedged Together

Nylon Plug exerts pressure here

Figure 1-106. Several types of automotive nuts (above). Cutaway view of one type of self-locking nut action.

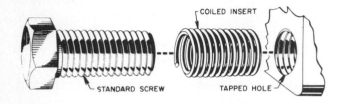

COILED INSERT

STANDARD SCREW   TAPPED HOLE

Figure 1-107. Damaged threads can be restored in a threaded part by the use of a helically coiled insert. The damaged hole is drilled to a precise oversize, tapped, and a coiled insert installed. This provides new threads of original diameter and type.

FLAT   COMMON LOCK   EXTERNAL LOCK   INTERNAL LOCK   EXTERNAL INTERNAL LOCK   COUNTER-SUNK LOCK

Figure 1-108. Common types of washers as used on the automobile. Lock washers prevent loosening of fasteners and should be used whenever original equipment was so equipped.

CONTRACTING

EXPANDING

Figure 1-109. Several types of commonly used snap rings. Snap rings are used to prevent endwise movement of shafts and bearings. Damaged or distorted snap rings must be replaced. *(Courtesy of Ford Motor Co. of Canada Ltd.)*

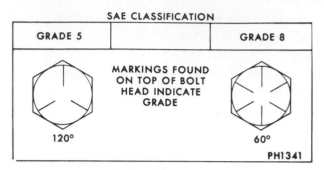

SAE CLASSIFICATION		
GRADE 5		GRADE 8
	MARKINGS FOUND ON TOP OF BOLT HEAD INDICATE GRADE	
120°		60°
		PH1341

BOLT TORQUE

Size	GRADE 5 Ft. Lbs.	newton metres	GRADE 8 Ft. Lbs.	newton metres
1/4-20	95 In. Lbs.	10.7	125 In. Lbs.	14.1
1/4-28	95 In. Lbs.	10.7	150 In. Lbs.	16.9
5/16-18	200 In. Lbs.	22.6	270 In. Lbs.	31.2
5/16-24	20	27.1	25	33.9
3/8-16	30	40.7	40	54.2
3/8-24	35	47.5	45	61.0
7/16-14	50	67.8	65	88.1
7/16-20	55	74.6	70	95.0
1/2-13	75	101.7	100	135.6
1/2-20	85	115.2	110	149.1
9/16-12	105	142.4	135	183.0
9/16-18	150	156.0	150	203.4
5/8-11	115	203.4	195	264.4
5/8-18	160	217.0	210	284.7
3/4-16	175	237.3	225	305.1

Figure 1-110. SAE bolt torque values in foot-pounds with metric Newton meter equivalents. Always adhere to the manufacturer's torque specifications. *(Courtesy of Chrysler Corporation)*

SHOP ROUTINE

TITLE: SAFETY

STUDENT'S NAME _____

PERFORMANCE OBJECTIVES

Study this chapter on personal safety, shop safety, and first aid in the automotive shop. After sufficient opportunity to study this portion of the text and the automotive shop with the instructor's supervision and demonstrations, you should be able to perform the following tasks at the request of the instructor.

TASK 1--- Use the proper clothing in the shop.
INSTRUCTOR CHECK_____

TASK 2--- Locate all fire extinguishers in the shop and state their proper use.
INSTRUCTOR CHECK_____

TASK 3--- Demonstrate the proper way to lift heavy objects.
INSTRUCTOR CHECK_____

TASK 4--- Demonstrate the proper use of shop exhaust ventilating equipment.
INSTRUCTOR CHECK_____

TASK 5--- Locate the first-aid station, identify its contents by name, and state the use of each item it contains.
INSTRUCTOR CHECK_____

TASK 6--- Locate the resident nurse's office and emergency telephone numbers (nurse, doctor, fire, ambulance, police).
INSTRUCTOR CHECK_____

PERFORMANCE EVALUATION

Your instructor may require you to perform these tasks in any of the following ways in order to evaluate your performance:
- By asking test questions
- By asking you to describe the performance of these tasks in writing
- By asking you to describe the performance of these tasks orally
- By asking you to perform these tasks when required

SHOP ROUTINE

**TITLE: SHOP MANUALS,
WORK ORDERS,
PARTS LIST,
TOOLS, AND EQUIPMENT**

STUDENT'S NAME _____

PERFORMANCE OBJECTIVES

Study this chapter on these titles. After sufficient opportunity to study this portion of the text and the use of shop manuals, work orders, parts lists, tools, and equipment with the instructor's supervision and demonstrations, you should be able to perform all the following tasks at the request of your instructor.

TASK 1---Demonstrate the use of a shop manual by locating specific procedures and specifications for a vehicle specified by your instructor.
INSTRUCTOR CHECK_____

TASK 2---Complete a work order with all the required information on a vehicle and repair job specified by your instructor.
INSTRUCTOR CHECK_____

TASK 3---Prepare a parts list with all the required information on a vehicle and repair job specified by your instructor.
INSTRUCTOR CHECK_____

TASK 4---Identify by name and demonstrate the proper use of all tools and equipment in this section of the text.
INSTRUCTOR CHECK_____

PERFORMANCE EVALUATION

Your instructor may require you to perform these tasks in any of the following ways in order to evaluate your performance:
- By asking test questions
- By asking you to describe the performance of these tasks in writing
- By asking you to describe the performance of these tasks orally
- By asking you to perform these tasks when required

STUDENT'S NAME _____

PERFORMANCE OBJECTIVES

Study this chapter on this title. After sufficient opportunity to study this portion of the text and the various types of fasteners with the instructor's supervision and demonstrations, you should be able to perform the following tasks at the request of your instructor.

TASK 1--- Identify the different types and sizes of automotive fasteners and explain their uses.
INSTRUCTOR CHECK_____

TASK 2--- Identify the different types of fastener threads.
INSTRUCTOR CHECK_____

TASK 3--- Identify the different tensile strenghts of fasteners.
INSTRUCTOR CHECK_____

PERFORMANCE EVALUATION

Your instructor may require you to perform these tasks in any of the following ways in order to evaluate your performance:

- By asking test questions
- By asking you to describe the performance of these tasks in writing
- By asking you to describe the performance of these tasks orally
- By asking you to perform these tasks when required

Self-Check

1. What type of shoes is best for automotive shop use?

2. Why is adequate ventilation of such critical importance in an automotive shop?

3. Make a list of as many safety rules as you can regarding the use of shop tools and equipment.

4. Why are shop manuals necessary in the automotive shop?

5. What information is necessary about a vehicle in order to obtain correct replacement parts for it?

6. What is torque?

7. Why is a torque wrench necessary?

8. A well-equipped technician will have a good range of customary size wrenches as well as _____ sizes.

9. To measure with an outside micrometer, the part being measured must be placed between the _____ and the _____.

10. List three safety rules that should be practiced when using power tools.

11. A hoist used to raise a car may be positioned to contact any point under the car. True or false?

12. What is the purpose of jack stands?

Performance Evaluation

Study this chapter and the proper use of the various tools and shop equipment used in the automotive shop. After enough opportunity to practice using the tools and equipment in a safe and efficient manner, you should be able to accomplish the following.

1. Recognize and practice safety in selecting and using proper clothing for work in an automotive shop.

2. Follow the required procedures in case of fire in the shop.

3. Use proper ventilation and shop exhaust equipment whenever needed.

4. Select and use the correct hand tools and power equipment in a safe and efficient manner.

5. Follow the first-aid procedures of the shop in which you are working.

6. Select the appropriate shop manual; locate and use the required information for the job being done.

7. Interpret work order information correctly.

8. Prepare a parts list with all the required information to obtain correct replacement.

9. Identify the correct types of gaskets used in repair work.

10. Select and use the correct type of fastener.

11. Complete the Self-Check with at least 80 percent accuracy.

12. Complete all practical work with 100 percent accuracy.

Chapter 2

Seals and Bearings

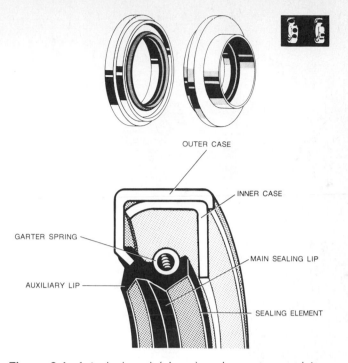

Figure 2-1. A typical seal (above) and common seal terminology (below). Seals are used to keep in oils, fluids, and grease, to exclude dirt, or both. The main sealing element can be synthetic rubber, leather, or felt. Some seals have both an inner and outer case; others have only an outer case. Some seals include a bolt-on flange.

Many different types of seals and bearings are used in the steering and suspension systems of the automobile. Seals are designed to prevent the entry of dirt and foreign material and to prevent the loss of grease, oil, or gases from the component. Bearings are designed to reduce friction and allow rotation of parts or the oscillation of parts. Bearings also allow for replacement of the bearing rather than replacing the entire component, which could be very costly in many cases.

PART 1   SEALS AND BEARINGS

Seals

Oil seals are classified as static or dynamic. The static seal is used between two stationary parts. The dynamic seal provides a seal between a stationary and a moving part. One example of a static seal is the steering gear cover gasket or 0-ring. The power steering pump shaft seal is an example of a dynamic seal.

Some seals are designed to withstand high pressures. The power steering gear rack seal is designed to withstand high hydraulic pressures and seals between two hydraulic chambers. Other seals use felt, synthetic rubber, fiber, or leather. Many seals have a metal case

and a tension spring. Both single- and double-lip seals are used.

Seals should always be installed according to the manufacturer's specification. In general, though, the sealing lip should be installed toward the fluid or gas being contained. Felt dust seals should be installed with the felt toward the outside. Dynamic seals should be lubricated in the sealing lip area when installed.

Bearings

Bearings reduce the friction of rotating parts. Overhauling and reconditioning are easier and less expensive when bearings can be replaced.

There are two basic classifications of bearings: friction and antifriction bearings. Friction bearings rely on sliding friction within a film of oil between the bearing and bearing journal. A friction bearing is a thin-walled cylindrical part located between the rotating part and the supporting part. It can be of one-piece construction or it may consist of several pieces. The one-piece friction bearing is often called a *bushing*, while the terms *bearing inserts* or *bearing shells* are used for

Figure 2-2. Several types of friction bearings and bushings. One-piece bushings are used on king pin solid axle suspension systems. Some bushings of a porous material are permanently lubricated with a special lubricant that saturates the bushings.

multiple-piece friction bearings. Various types and combinations of metals are used for friction bearings. Sintered bronze, brass, steel, babbitt metal, lead, tin, nylon, and aluminum are examples. There are no balls or rollers in friction-type bearings.

Antifriction bearings rely on rolling friction for operation. Rolling friction requires less effort to rotate a part than does sliding friction. The greater the load, the more evident this fact becomes.

Antifriction bearings include various designs of ball bearings and roller bearings. Bearings are designed to carry radial loads, axial loads, or both. A radial load is a load imposed at right angles to the shaft. An axial load is a load imposed parallel to the shaft.

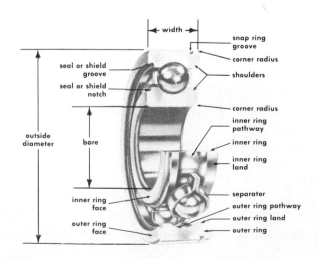

Figure 2-4. Tapered roller bearing and component part names. This bearing is able to carry a radial load as well as an axial load in one direction. *(Courtesy of General Motors Corporation)*

A straight roller bearing would be able to carry a radial load, whereas a thrust bearing would be able to absorb an axial load. A tapered roller bearing could carry both a radial load and a unidirectional axial load. A double-row, opposed, tapered roller bearing is able to carry a radial load as well as axial loads in both directions.

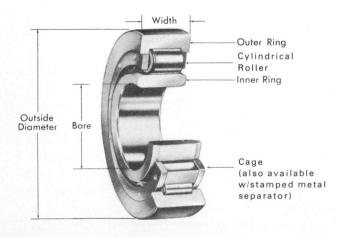

Figure 2-3. Cutaway view of a straight roller bearing with parts identified. This bearing is designed to carry a radial load (90 degrees to the shaft).

Figure 2-5. Cutaway view of a ball bearing with parts identified. This bearing is also designed to carry a radial load.

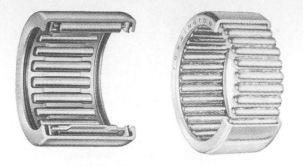

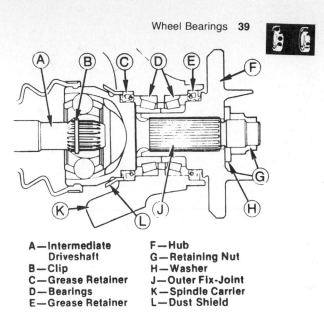

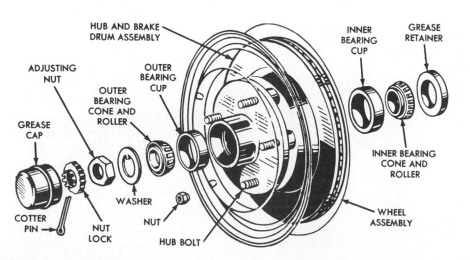

Figure 2-6. Common types of needle bearings; designed to carry radial loads.

A—Intermediate Driveshaft	F—Hub
B—Clip	G—Retaining Nut
C—Grease Retainer	H—Washer
D—Bearings	J—Outer Fix-Joint
E—Grease Retainer	K—Spindle Carrier
	L—Dust Shield

Figure 2-8. Nonadjustable tapered roller wheel bearing arrangement on one front-wheel-drive model. *(Courtesy of Ford Motor Co.)*

PART 2   WHEEL BEARINGS

There are several types of wheel bearing arrangements for both driving and nondriving wheels, whether front-wheel drive, rear-wheel drive, or four-wheel drive. One design uses two opposed, tapered roller bearings with the inner bearing being the larger of the two. This arrangement usually provides for replacement of faulty bearings and allows periodic repacking of the bearings. In many applications of this design, a precise adjustment of the wheel bearings is required, while on other applications adjustment is automatically provided by tightening the retaining nut to specified torque.

Another design uses a preassembled, prelubricated, and sealed bearing and hub assembly. This design does not provide for periodic repacking of wheel bearings; no wheel-bearing adjustment is required or possible. A wheel-bearing problem requires replacement of the bearing and hub assembly.

A variation of the sealed type allows replacement of the sealed bearing without replacing the hub. The sealed unit may have roller bearings or ball bearings.

Figure 2-7. Exploded view of wheel hub and bearing assembly. *(Courtesy of Ford Motor Co. of Canada Ltd.)*

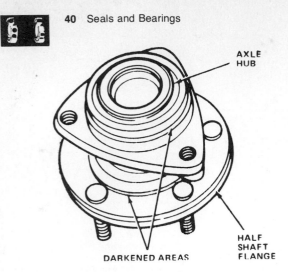

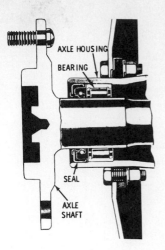

Figure 2-9. The American Motors four-wheel-drive Eagle models have a unique front axle hub-and-bearing assembly. The assembly is sealed and does not require lubrication or periodic maintenance. The hub has ball bearings that seat in races machined directly into the hub. There are darkened areas surrounding the bearing face areas of the hub. These darkened areas are from a heat treatment process, are normal, and should not be mistaken for a problem condition. *(Courtesy of American Motors Corporation)*

Figure 2-11. Straight roller bearing rear-axle bearing-and-seal arrangement. Notice the absence of a separate inner bearing race. Rollers run on axle shaft. *(Courtesy of General Motors Corporation)*

Many rear-wheel-drive axle bearings are of the straight roller bearing design. These usually use the drive axle shaft as the inner bearing race. Other rear-wheel-drive axle bearings are of the tapered roller bearing or ball bearing type. In these types a complete bearing is used rather than having the axle shaft serve as the inner race. Lubrication is provided from the differential lubricant in some cases, while other designs may require packing the bearing with wheel-bearing grease or multipurpose grease during assembly or a prelubricated, sealed bearing may be used.

For further details, see the wheel bearing diagnosis and service section following.

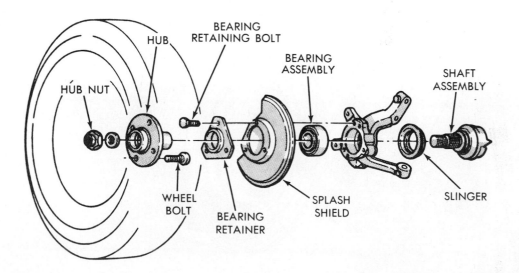

Figure 2-10. Sealed bearing type of front-wheel-drive and mounting arrangement. *(Courtesy of Chrysler Corporation)*

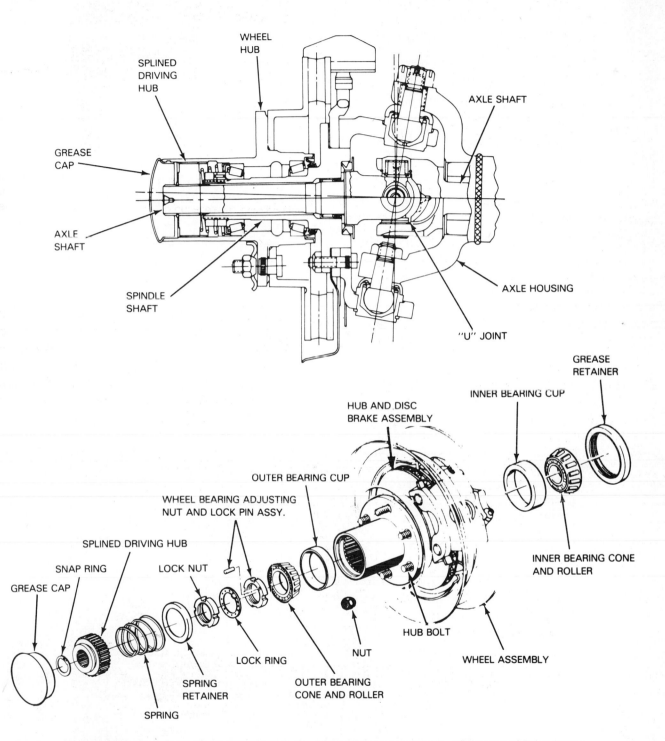

SPLINED
DRIVING
HUB

WHEEL
HUB

AXLE SHAFT

GREASE
CAP

AXLE
SHAFT

SPINDLE
SHAFT

"U" JOINT

AXLE HOUSING

GREASE
RETAINER

INNER BEARING CUP

HUB AND DISC
BRAKE ASSEMBLY

OUTER BEARING CUP

WHEEL BEARING ADJUSTING
NUT AND LOCK PIN ASSY.

SPLINED DRIVING HUB

SNAP RING

LOCK NUT

GREASE CAP

SPRING
RETAINER

SPRING

LOCK RING

OUTER BEARING
CONE AND ROLLER

NUT

HUB BOLT

WHEEL ASSEMBLY

INNER BEARING CONE
AND ROLLER

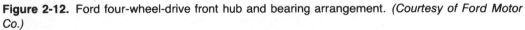

Figure 2-12. Ford four-wheel-drive front hub and bearing arrangement. *(Courtesy of Ford Motor Co.)*

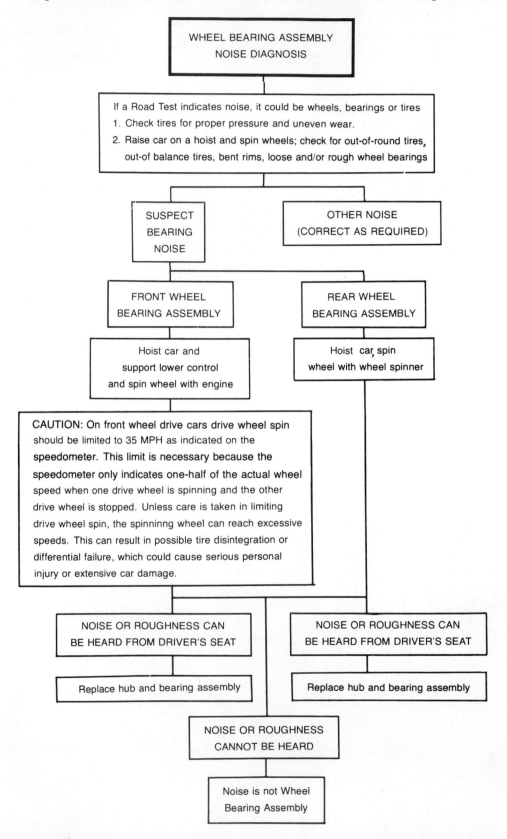

Wheel bearing noise diagnosis procedure for Citation front-wheel-drive vehicle. In general the same procedure may be used for most drive-wheel noise problems front or rear. (Courtesy of General Motors Corporation)

WHEEL BEARING ASSEMBLY
NOISE DIAGNOSIS

If a Road Test indicates noise, it could be wheels, bearings or tires
1. Check tires for proper pressure and uneven wear.
2. Raise car on a hoist and spin wheels; check for out-of-round tires, out-of balance tires, bent rims, loose and/or rough wheel bearings

SUSPECT
BEARING
NOISE

OTHER NOISE
(CORRECT AS REQUIRED)

FRONT WHEEL
BEARING ASSEMBLY

REAR WHEEL
BEARING ASSEMBLY

Hoist car and
support lower control
and spin wheel with engine

Hoist car, spin
wheel with wheel spinner

CAUTION: On front wheel drive cars drive wheel spin should be limited to 35 MPH as indicated on the speedometer. This limit is necessary because the speedometer only indicates one-half of the actual wheel speed when one drive wheel is spinning and the other drive wheel is stopped. Unless care is taken in limiting drive wheel spin, the spinninng wheel can reach excessive speeds. This can result in possible tire disintegration or differential failure, which could cause serious personal injury or extensive car damage.

NOISE OR ROUGHNESS CAN
BE HEARD FROM DRIVER'S SEAT

NOISE OR ROUGHNESS CAN
BE HEARD FROM DRIVER'S SEAT

Replace hub and bearing assembly

Replace hub and bearing assembly

NOISE OR ROUGHNESS
CANNOT BE HEARD

Noise is not Wheel
Bearing Assembly

Typical sealed wheel bearing looseness diagnosis procedure for General Motors Citation. (Courtesy of General Motors Corporation)

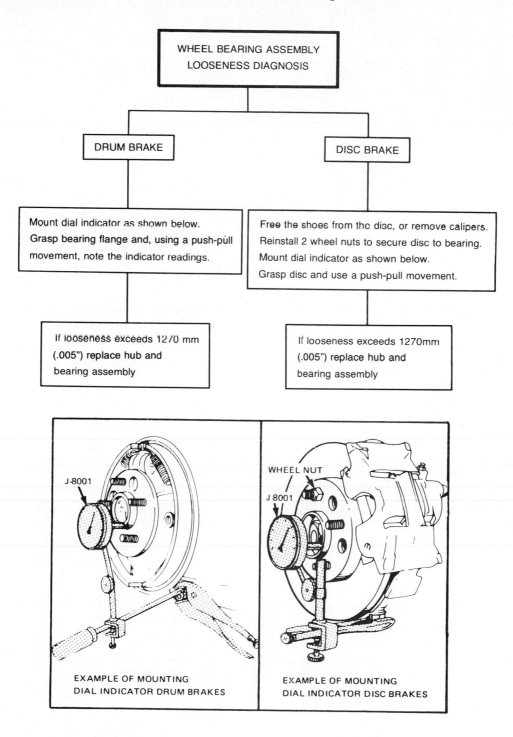

WHEEL BEARING ASSEMBLY
LOOSENESS DIAGNOSIS

DRUM BRAKE

DISC BRAKE

Mount dial indicator as shown below.
Grasp bearing flange and, using a push-pull
movement, note the indicator readings.

Free the shoes from the disc, or remove calipers.
Reinstall 2 wheel nuts to secure disc to bearing.
Mount dial indicator as shown below.
Grasp disc and use a push-pull movement.

If looseness exceeds 1270 mm
(.005") replace hub and
bearing assembly

If looseness exceeds 1270mm
(.005") replace hub and
bearing assembly

J-8001

EXAMPLE OF MOUNTING
DIAL INDICATOR DRUM BRAKES

WHEEL NUT

J 8001

EXAMPLE OF MOUNTING
DIAL INDICATOR DISC BRAKES

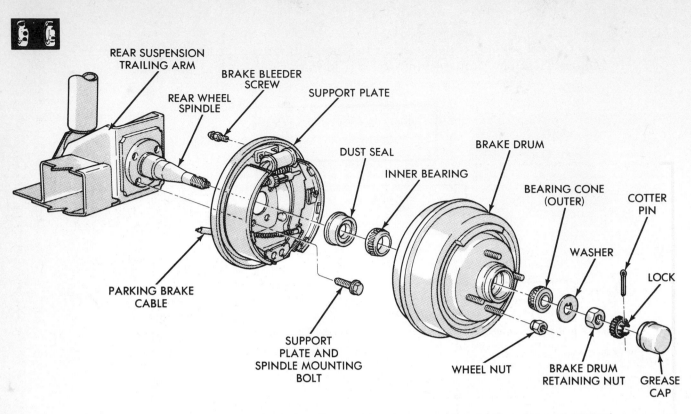

Figure 2-13. Trailing rear-axle wheel mounting and bearing arrangement. Two opposed tapered roller bearings are used, and bearings are adjustable. *(Courtesy of Chrysler Corporation)*

General Precautions

• Raise the vehicle only at the contact points specified by the vehicle manufacturer.

• The vehicle must be properly supported at all times.

• Use only the proper tools and supports in good condition and use them as recommended.

• Do not disassemble any suspension parts without proper consideration of the consequences; a loaded spring packs a powerful punch.

• Always refer to the manufacturer's manual for proper procedures; then follow them.

• Never allow a wheel-and-tire assembly to bounce; keep the wheel under control at all times.

• Never depress the brake pedal when brake parts have been removed.

• Never mix parts from one wheel assembly with another.

• Never add grease to wheel bearings. Completely remove all old grease and use only new grease.

• Never use suspension parts that have been heated, damaged, bent, or straightened.

• Avoid getting grease, solvent, brake fluid, or dirty fingerprints on brake discs, pads, linings, and drums.

• Never allow suspension or brake parts to hang on brake hoses or constant velocity universal joints.

• Always use the correct size and type of fastener.

• Use new fasteners when recommended by the manufacturer.

• Always tighten all fasteners to specified torque.

• Always install new cotter pins properly locked wherever needed.

WHEEL BEARING DIAGNOSIS

CONSIDER THE FOLLOWING FACTORS WHEN DIAGNOSING BEARING CONDITION:

1. GENERAL CONDITION OF ALL PARTS DURING DISASSEMBLY AND INSPECTION.
2. CLASSIFY THE FAILURE WITH THE AID OF THE ILLUSTRATIONS.
3. DETERMINE THE CAUSE.
4. MAKE ALL REPAIRS FOLLOWING RECOMMENDED PROCEDURES.

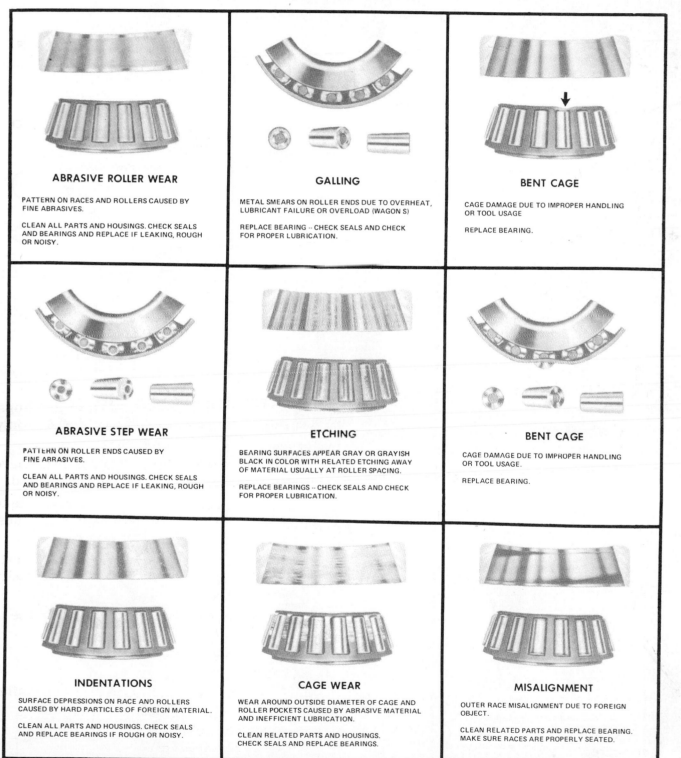

ABRASIVE ROLLER WEAR

PATTERN ON RACES AND ROLLERS CAUSED BY FINE ABRASIVES.

CLEAN ALL PARTS AND HOUSINGS. CHECK SEALS AND BEARINGS AND REPLACE IF LEAKING, ROUGH OR NOISY.

GALLING

METAL SMEARS ON ROLLER ENDS DUE TO OVERHEAT, LUBRICANT FAILURE OR OVERLOAD (WAGON S)

REPLACE BEARING -- CHECK SEALS AND CHECK FOR PROPER LUBRICATION.

BENT CAGE

CAGE DAMAGE DUE TO IMPROPER HANDLING OR TOOL USAGE

REPLACE BEARING.

ABRASIVE STEP WEAR

PATTERN ON ROLLER ENDS CAUSED BY FINE ABRASIVES.

CLEAN ALL PARTS AND HOUSINGS. CHECK SEALS AND BEARINGS AND REPLACE IF LEAKING, ROUGH OR NOISY.

ETCHING

BEARING SURFACES APPEAR GRAY OR GRAYISH BLACK IN COLOR WITH RELATED ETCHING AWAY OF MATERIAL USUALLY AT ROLLER SPACING.

REPLACE BEARINGS -- CHECK SEALS AND CHECK FOR PROPER LUBRICATION.

BENT CAGE

CAGE DAMAGE DUE TO IMPROPER HANDLING OR TOOL USAGE.

REPLACE BEARING.

INDENTATIONS

SURFACE DEPRESSIONS ON RACE AND ROLLERS CAUSED BY HARD PARTICLES OF FOREIGN MATERIAL.

CLEAN ALL PARTS AND HOUSINGS. CHECK SEALS AND REPLACE BEARINGS IF ROUGH OR NOISY.

CAGE WEAR

WEAR AROUND OUTSIDE DIAMETER OF CAGE AND ROLLER POCKETS CAUSED BY ABRASIVE MATERIAL AND INEFFICIENT LUBRICATION.

CLEAN RELATED PARTS AND HOUSINGS. CHECK SEALS AND REPLACE BEARINGS.

MISALIGNMENT

OUTER RACE MISALIGNMENT DUE TO FOREIGN OBJECT.

CLEAN RELATED PARTS AND REPLACE BEARING. MAKE SURE RACES ARE PROPERLY SEATED.

(Courtesy of General Motors Corp.)

REAR AXLE BEARING DIAGNOSIS

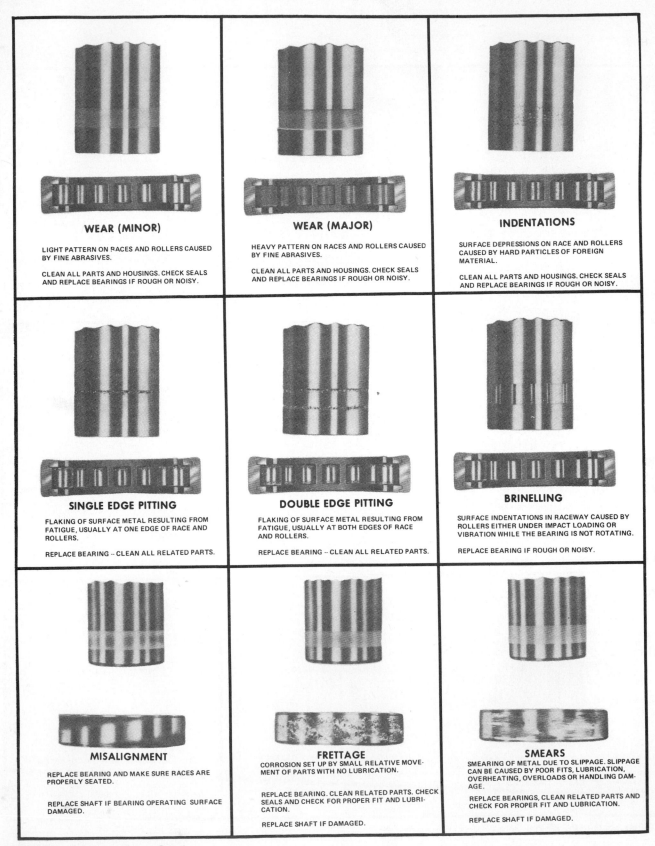

WEAR (MINOR)

LIGHT PATTERN ON RACES AND ROLLERS CAUSED BY FINE ABRASIVES.

CLEAN ALL PARTS AND HOUSINGS. CHECK SEALS AND REPLACE BEARINGS IF ROUGH OR NOISY.

WEAR (MAJOR)

HEAVY PATTERN ON RACES AND ROLLERS CAUSED BY FINE ABRASIVES.

CLEAN ALL PARTS AND HOUSINGS. CHECK SEALS AND REPLACE BEARINGS IF ROUGH OR NOISY.

INDENTATIONS

SURFACE DEPRESSIONS ON RACE AND ROLLERS CAUSED BY HARD PARTICLES OF FOREIGN MATERIAL.

CLEAN ALL PARTS AND HOUSINGS. CHECK SEALS AND REPLACE BEARINGS IF ROUGH OR NOISY.

SINGLE EDGE PITTING

FLAKING OF SURFACE METAL RESULTING FROM FATIGUE, USUALLY AT ONE EDGE OF RACE AND ROLLERS.

REPLACE BEARING -- CLEAN ALL RELATED PARTS.

DOUBLE EDGE PITTING

FLAKING OF SURFACE METAL RESULTING FROM FATIGUE, USUALLY AT BOTH EDGES OF RACE AND ROLLERS.

REPLACE BEARING -- CLEAN ALL RELATED PARTS.

BRINELLING

SURFACE INDENTATIONS IN RACEWAY CAUSED BY ROLLERS EITHER UNDER IMPACT LOADING OR VIBRATION WHILE THE BEARING IS NOT ROTATING.

REPLACE BEARING IF ROUGH OR NOISY.

MISALIGNMENT

REPLACE BEARING AND MAKE SURE RACES ARE PROPERLY SEATED.

REPLACE SHAFT IF BEARING OPERATING SURFACE DAMAGED.

FRETTAGE

CORROSION SET UP BY SMALL RELATIVE MOVEMENT OF PARTS WITH NO LUBRICATION.

REPLACE BEARING. CLEAN RELATED PARTS. CHECK SEALS AND CHECK FOR PROPER FIT AND LUBRICATION.

REPLACE SHAFT IF DAMAGED.

SMEARS

SMEARING OF METAL DUE TO SLIPPAGE. SLIPPAGE CAN BE CAUSED BY POOR FITS, LUBRICATION, OVERHEATING, OVERLOADS OR HANDLING DAMAGE.

REPLACE BEARINGS, CLEAN RELATED PARTS AND CHECK FOR PROPER FIT AND LUBRICATION.

REPLACE SHAFT IF DAMAGED.

(Courtesy of General Motors Corp.)

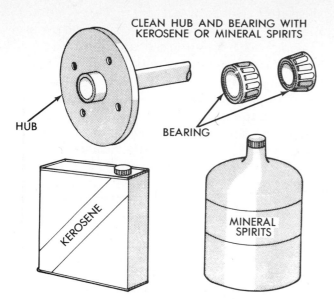

CLEAN HUB AND BEARING WITH
KEROSENE OR MINERAL SPIRITS

HUB

BEARING

KEROSENE

MINERAL
SPIRITS

Figure 2-14. *(Courtesy of Chrysler Corporation)*

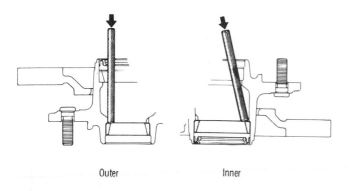

Outer

Inner

Figure 2-15. Removing wheel bearing outer races or cups from front-wheel hub on rear-wheel-drive vehicle. *(Courtesy of Chrysler Corporation)*

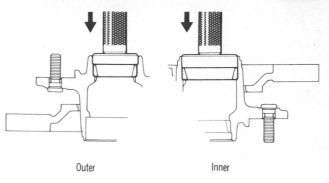

Outer

Inner

Figure 2-16. Using bearing driver to install front-wheel bearing cups into front-wheel hubs. Cups must be fully seated. *(Courtesy of Chrysler Corporation)*

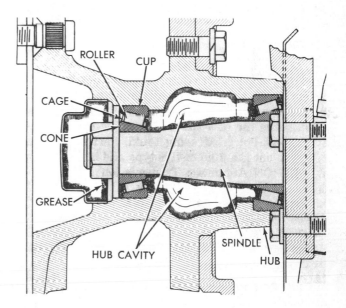

ROLLER   CUP

CAGE

CONE

GREASE

HUB CAVITY

SPINDLE

HUB

Figure 2-17. Dark shaded area shows grease in hub, bearings, spindle, and dust cap on properly lubricated front wheel bearings. *(Courtesy of Chrysler Corporation)*

Wheel Bearing Removal

Front or Rear

The procedure for removal of wheel bearings will vary considerably depending on the following design differences:

- Front wheel, nondriving, disc brakes
- Front wheel, nondriving, drum brakes
- Front wheel, driving, drum brakes
- Rear wheel, nondriving, drum brakes
- Rear wheel, driving, drum brakes
- Rear wheel, driving, disc brakes

First position the vehicle in the shop where it is going to be serviced. Next remove the wheel covers and loosen wheel-retaining nuts one turn before raising the vehicle. *Caution* Never move or drive a vehicle with wheel nuts loosened; damage to wheels will result.

Raise the vehicle on a frame contact hoist to allow the wheels to hang from the suspension or jack the vehicle and place safety stands under the vehicle to achieve the same result. Mark one wheel stud and wheel stud hole to maintain wheel balance after assembly by maintaining original wheel-to-hub position. Chalk mark all wheels as to position on the vehicle---left front, right front, left rear, right rear---for proper reassembly.

Remove the wheels. On drum brake units mark the brake drums similarly, then remove the drums. This may require retracting the brake shoe adjustment. This can be done by holding the self-adjuster away from the star wheel adjuster and turning the adjuster in the correct direction. Access to the self-adjuster is made through a hole in the backing plate in most cases. On ratchet-type self-adjusters, pry the spring-loaded ratchet back to the release position.

On rear-wheel drum units the drum may be held by speed nut retainers on the wheel-mounting studs or by screws that hold the drum to the axle flange. These must first be removed to allow drum removal. Handle drums with care to prevent distortion. Do not use a hammer on the brake drums. If drums are tight, use the drum puller recommended for the purpose. However, incorrect use of the puller, such as applying too much force, can also distort the drum.

On disc brake units remove the caliper from the steering knuckle and support the caliper so that it will not hang by the flexible brake hose.

On many front-wheel units and some rear-wheel units, the dust cap must be removed to allow access to the retaining nut. The dust cap is a tight press fit into the wheel hub and should be removed with a special tool or pried off with a screwdriver to prevent damage to the dust cap.

Next remove the cotter pin by using a pair of side cutters. Grip the cotter pin head firmly with the side cutters and pry up. If the nut is staked to a groove in the spindle or axle shaft, carefully force the staked part of the nut up out of the groove with a special tapered narrow chisel-type tool. Lubricate the threads if dry and remove the nut. In most cases this will allow the removal of the retaining washer and hub assembly from the spindle.

Next pry the inner seal from the hub and remove the bearing cone. Do the same with the outer seal if so equipped.

Repacking and Assembly

Wipe all excess grease from the spindle, hub, and bearings; then wash these parts in cleaning solvent. These parts must be absolutely clean of all old grease and dirt. Blow dry with compressed air but never spin bearings with compressed air. Spinning will damage the bearing; it may fly apart, causing injury and damage.

Examine all parts for wear or damage. If bearing cups in the hub are tight and in good condition, they may be used again. If not, replace them as shown in the illustrations.

Repack good used bearing cones with the recommended grade of wheel bearing grease or multipurpose grease. Make sure that all rollers are completely surrounded with grease. This can be done by hand or with a bearing packer designed for the purpose. Also, apply a small layer of the same grease to the inside of the hub, to the spindle, and to the dust cap. This prevents rust and provides a reserve of lubricant.

Install the inner wheel-bearing cone and new dust seal in the hub. Position the hub over the spindle without damaging the seal and install the outer bearing washer and retaining nut.

Follow the manufacturer's recommended procedure in the service manual for correct tightening of the nut. The general procedure for the adjustable type is to tighten the nut to specified torque to seat the bearings. Then back the nut off one turn. Next tighten the nut either finger-tight or to specified torque. This provides the necessary end play (no preload) for the bearings. Next install the cotter pin in the castellated type of nut. If the holes do not line up, the nut should be backed off to the first hole only; then install the cotter pin of the correct size and lock it securely. If a plain nut and nut lock are used, position the nut lock so that the cotter pin holes line up; install and lock the cotter pin. Now install the dust cap without distorting or denting it.

On drum brake units, preliminary adjustment of brake shoes may be required before drum installation (see the brake service section in this text).

Install the brake caliper or brake drum (depending on design) and tighten all bolts to specifications. Install the wheels, taking note of the position markings, and tighten lug nuts in the proper sequence and to proper torque. This is critical to prevent disc or drum distortion and to assure lug nuts do not loosen. Undertightening and overtightening

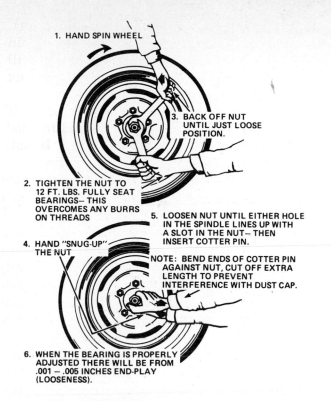

1. HAND SPIN WHEEL

3. BACK OFF NUT UNTIL JUST LOOSE POSITION.

2. TIGHTEN THE NUT TO 12 FT. LBS. FULLY SEAT BEARINGS— THIS OVERCOMES ANY BURRS ON THREADS

5. LOOSEN NUT UNTIL EITHER HOLE IN THE SPINDLE LINES UP WITH A SLOT IN THE NUT— THEN INSERT COTTER PIN.

4. HAND "SNUG-UP" THE NUT

NOTE: BEND ENDS OF COTTER PIN AGAINST NUT, CUT OFF EXTRA LENGTH TO PREVENT INTERFERENCE WITH DUST CAP.

6. WHEN THE BEARING IS PROPERLY ADJUSTED THERE WILL BE FROM .001 – .005 INCHES END-PLAY (LOOSENESS).

Figure 2-18. Typical wheel bearing adjustment procedure for one type of bearing arrangement. Procedures and specifications vary depending on design. Manufacturer's specifications and procedures must be followed. *(Courtesy of General Motors Corporation)*

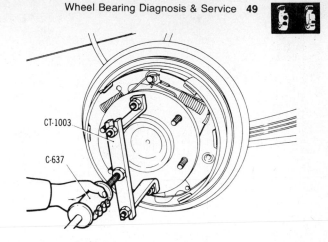

CT-1003

C-637

Figure 2-20. Removing rear drive axle shaft with a slide hammer type of puller after axle shaft retainer bolts or C-lock have been removed. *(Courtesy of Chrysler Corporation)*

are equally undesirable. Depress the brake pedal several times to restore pedal reserve after all brake parts are reassembled.

Replacing Sealed Wheel Bearings

Front or Rear

Many front-wheel-drive vehicles have the sealed type of wheel bearings, which do not require periodic service or repacking with

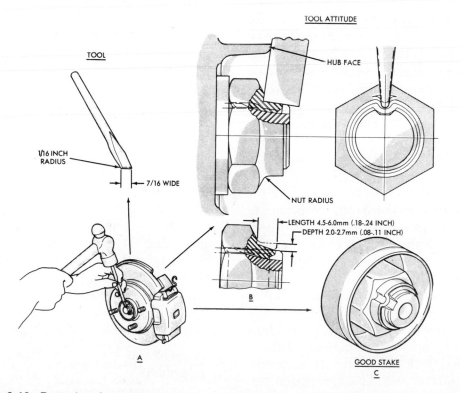

Figure 2-19. Procedure for staking (locking) one type of front-wheel bearing retaining nut. *(Courtesy of Chrysler Corporation)*

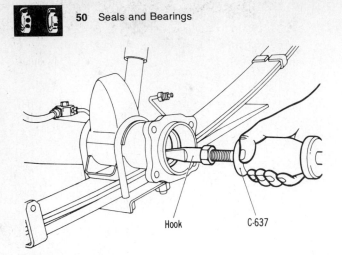

Figure 2-21. Removing inner axle shaft seal from axle housing with a slide hammer type of seal puller. *(Courtesy of Chrysler Corporation)*

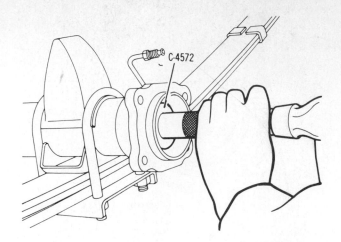

Figure 2-22. Installing inner axle shaft seal with proper seal driver. Seal must be installed proper side out and to correct depth in housing. Refer to shop manual for specifications. *(Courtesy of Chrysler Corporation)*

grease. These bearings are serviced only when a wheel bearing noise problem or roughness problem occurs, at which time the old bearings are removed and new bearings installed.

Disassembly of disc and drum brake units on these vehicles is similar to that of other repackable bearing designs. In some cases the hub-and-bearing assembly must be replaced as a unit. In others the sealed bearing can be replaced separately.

Most sealed bearing arrangements do not require any bearing adjustment other than tightening the retaining nut to specified torque and locking the nut either by staking or with a cotter pin. The manufacturer's recommended procedures and specifications must be followed when replacing sealed bearings.

Rear Axle Bearings and Seals

Rear-Wheel Drive

When a rear-axle bearing problem is suspected, a thorough diagnosis must be performed to determine the problem and identify the particular axle bearing causing the problem. Follow the problem diagnosis procedure at the beginning of this part and the vehicle manufacturer's diagnosis procedure.

Once the problem bearing has been identified, the bearing must be replaced. Follow the recommended jacking or hoisting procedure and use safety stands to support the vehicle. Remove the wheel after proper marking of the wheel and stud and proceed to disassemble the disc or drum brake unit as outlined earlier for wheel bearing service.

Rear-drive axle shafts are retained in the axle housing in one of several ways. Some are held in place by the axle bearing and a retainer plate at the outer end of the axle housing. Others are held in place by a C-shaped lock that fits into a groove on the axle shaft at the inner end of the shaft, inside the differential. This latter type requires draining the differential gear oil and removing the differential inspection cover. This provides access to remove the differential pinion gear shaft retaining screw or pin and pinion gear shaft. Removing the pinion gear shaft (sometimes called spider gear shaft) allows the axle shaft to be pushed in to expose the C-lock for removal.

The retainer plate at the outer end should then be removed. Retainer bolts are accessible through the hole in the axle drive flange. The axle shaft can then be removed with the appropriate slide hammer puller as illustrated.

There are two basic types of bearing mounting. One type has the complete bearing mounted by a press fit on the axle shaft and retained by a press fit retaining ring. The other type has the bearing mounted in the axle housing and the axle shaft machined surface serves as the inner bearing race.

Remove and Install Axle Bearing and Seal (from axle shaft)

The rear-wheel bearing and bearing retainer ring both have a heavy press fit on the axle shaft. Because of this fit, they should be removed or installed separately. Both the retainer ring and the bearing must be removed to replace the seal.

1. Position and tighten the axle shaft in a vise at an angle so that the retainer ring rests on the vise jaws. Use a heavy chisel and hammer to crack the retainer ring. *Do not use heat to remove the retainer ring, as this may temper the axle shaft and result in axle shaft failure.*

2. Press the axle bearing off, using the recommended tools and equipment. Follow the equipment manufacturer's directions for setting up the axle and press to avoid damage to parts and equipment and to avoid injury to yourself and others.

3. Remove the axle shaft seal and retainer plate.

4. If necessary, install a new retainer plate and seal on the axle shaft, then install the bearing.

5. Press the bearing into place, using the recommended tools and equipment and following the manufacture's instructions for setting up and operating the equipment. Make sure that the bearing is pressed fully against the shoulder or on the axle shaft.

6. Press the new bearing retainer on the axle shaft with the proper tool and make sure that the retainer is seated against the bearing.

7. If the axle housing has an inner seal, remove and replace it with a new seal. On this

type, the axle bearing is either a sealed bearing or must be packed with wheel bearing grease.

8. Install the axle shaft in the housing and install and tighten the retaining plate bolts and nuts to specified torque.

9. Push the axle shaft in all the way and install the C-locks, pinion shaft, and lock if so equipped. Clean the differential inspection cover and mounting surface on the axle housing. Do not allow any foreign matter to enter the differential housing. Apply new gasket or RTV (room temperature vulcanizing) sealer to cover as recommended and install cover and bolts. Tighten bolts to specified torque and fill the differential to the proper level with the lubricant type specified by the vehicle manufacturer.

10. Install the drum or disc and wheel assembly as outlined earlier and tighten all bolts to specified torque. Damaged bolts and nuts should not be used.

Remove and Install Bearing and Seal (from axle housing)

1. With the axle shaft removed, use a slide hammer puller to remove the seal and bearing from the axle housing.

2. Install a new bearing in the housing using the proper bearing driver. Make sure that the bearing is installed to the proper depth in the housing. Lubricate the bearing with wheel bearing grease.

3. Install a new seal with the sealing

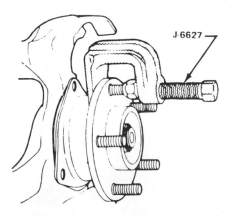

REMOVE WHEEL STUD

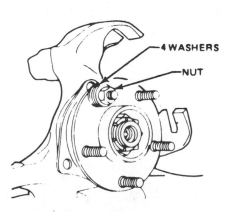

INSTALL WHEEL STUD

Figure 2-23. Removing and installing wheel mounting bolts. This method avoids damage to drive flange. *(Courtesy of General Motors Corporation)*

lip facing inward, using the proper size seal driver. Do not allow the seal to cock in the housing bore since this will damage the seal. Make sure that the seal is installed to the correct depth. Follow the manufacturer's instructions for this.

4. Lubricate the seal with wheel bearing grease.

5. Install the axle shaft (if in good condition), being careful not to damage the seal. Axle bearing and seal surfaces must be in good condition, the shaft must not be bent, splines must be in good condition, and the wheel mounting flange should not be bent. Install the retaining plate bolts and nuts and tighten to specified torque.

6. Push the axle shaft in all the way and install the C-locks, pinion shaft, and lock if so equipped. Clean the differential inspection cover and mounting surface on the axle housing. Do not allow an foreign matter to enter the differential housing. Apply new gasket or RTV sealer to cover as recommended and install cover and bolts. Tighten bolts to specified torque and fill the differential to the proper level with the lubricant type specified by the vehicle manufacturer.

7. Install the drum or disc and wheel assembly as outlined earlier and tighten all bolts to specified torque. Damaged bolts and nuts should not be used.

PART 4   WHEEL LUG BOLT REPLACEMENT

On flanged drive axles and on driving or nondriving wheel hubs where the lug bolts or studs are not swaged, the stud is removed with a C-clamp type of tool. Use of this type of tool prevents damage to the flange. The lug bolt should not be driven out with a hammer since this could distort the flange and cause excessive flange and wheel run out.

To install a new lug bolt place the bolt in the hole in the flange. Make sure that the serrations on the bolt and in the hole are aligned. Place enough flat washers on the bolt to cover half the bolt threads. Thread standard wheel nut on the lug bolt with the flat side of the nut against the washers. Tighten the nut until the bolt head is seated against the flange. Remove the nut and washers.

Swaged wheel studs must have the swaged stud material removed with the proper cutting tool before the stud is removed. The new stud is then installed and swaged with a special peening tool. Care must be exercised during this procedure not to damage the stud threads. On drum brake units the brake drum should be checked on a brake drum lathe to ensure drum concentricity. At least one light cut should be taken to determine concentricity. Refer to the section on brake system service for this procedure.

Chapter 2

PROJECT NO. 2-A

SEALS AND BEARINGS

TITLE: <u>SEALS AND BEARINGS</u>

STUDENT'S NAME _____

PERFORMANCE OBJECTIVES

Study this chapter. After sufficient opportunity to study this portion of the text and the various types of seals and bearings with the instructor's supervision and demonstrations, you should be able to perform the following tasks at the request of your instructor.

TASK 1---Identify the different types of seals.
INSTRUCTOR CHECK_____

TASK 2---Identify the different types of friction and anti-friction bearings and bearing materials and explain their uses.
INSTRUCTOR CHECK_____

TASK 3---Diagnose and correct front- and rear-wheel bearing problems.
INSTRUCTOR CHECK_____

PERFORMANCE EVALUATION

Your instructor may require you to perform these tasks in any of the following ways in order to evaluate your performance:
• By asking test questions
• By asking you to describe the performance of these tasks in writing
• By asking you to describe the performance of these tasks orally
• By asking you to perform these tasks when required

Self-Check

1. Name the two general classifications of seals.

2. What is the purpose of a front-wheel bearing seal?

3. Name three types of sealing materials.

4. Name the two general classifications of bearings.

5. List four different types of wheel and axle bearings.

6. Why is it necessary to mark wheels and wheel studs before wheel removal?

7. Describe the basic method of cleaning and repacking nondriving front- or rear-wheel bearings.

8. What is the proper procedure for wheel bearing adjustment?

9. Describe the proper method for tightening wheel mounting nuts.

10. To remove a damaged wheel lug bolt, just drive it out with a hammer. True or false?

11. When installing a new wheel lug, the serrations should be lined up. True or false?

12. Wheel lug bolts or nuts are all right-hand thread. True or false?

Performance Evaluation

After studying this chapter and with enough opportunity to practice the service procedures, and with the necessary tools, equipment, and supplies, you should be able to do the following.

1. State the purpose of seals and bearings.

2. Describe the basic construction and operation of seals and bearings including driving and nondriving wheel and axle bearings.

3. Diagnose, remove, clean, inspect, repack, replace, and adjust seals and bearings in driving and nondriving wheels and axles.

4. Prepare the vehicle for customer acceptance after completing the service.

5. Complete the Self-Check with at least 80 percent accuracy.

6. Complete all practical work with 100 percent accuracy.

Chapter 3

Brakes

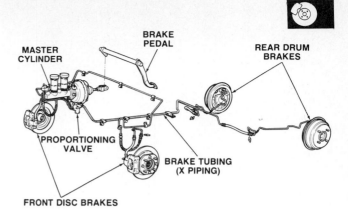

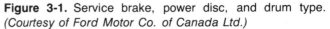

Figure 3-1. Service brake, power disc, and drum type. *(Courtesy of Ford Motor Co. of Canada Ltd.)*

The brake system is one of the most important safety systems on the automobile. The ability of the brake system to bring a vehicle to a safe controlled stop is absolutely essential in preventing accidental vehicle damage, personal injury, and loss of life.

To be able to identify and correct a brake system problem, to be able to restore a brake system to its maximum effectiveness, requires considerable knowledge of the system's construction and operation. This includes the friction devices at each of the vehicle's wheels and the mechanical and hydraulic control systems that control the action of these friction devices. A good basic understanding of the principles of mechanical devices, hydraulic systems, and friction devices is essential.

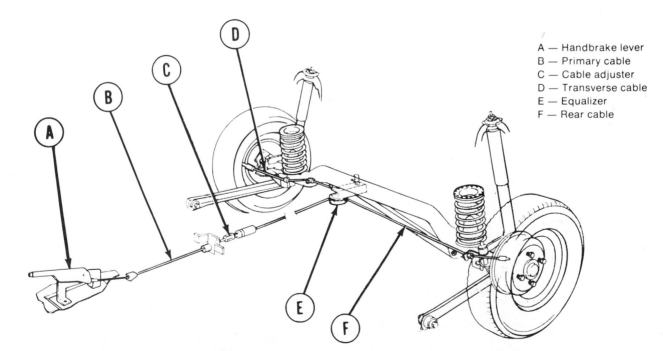

A — Handbrake lever
B — Primary cable
C — Cable adjuster
D — Transverse cable
E — Equalizer
F — Rear cable

Figure 3-2. Parking brake control mechanism for one make of front-wheel drive vehicle. *(Courtesy of Ford Motor Co. of Canada Ltd.)*

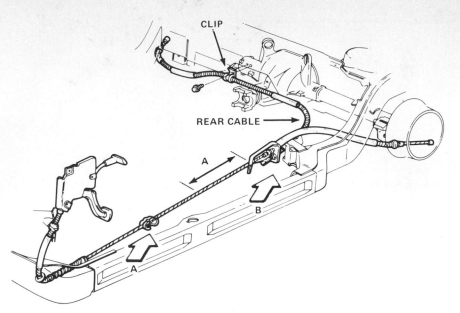

CLIP

REAR CABLE

A

B

A

Figure 3-3. Parking brake, foot-operated. *(Courtesy of General Motors Corporation)*

PART 1 HYDRAULIC AND BRAKING PRINCIPLES

Pressure, Force, and Motion

A hydraulic system uses a liquid to transmit pressure, force, and motion. When force is applied to a confined liquid, a pressure is produced and exerted undiminished throughout the system. This pressure acts at right angles to all surfaces in the system and with equal force on equal areas. This is known as Pascal's law and is basic to all hydraulic systems.

Pressure is stated in pounds per square inch or in kilopascals. Force is the amount of force produced by the output piston as the input piston is moved. If there is only one input piston and only one output piston, both with the same reaction area, they will travel an equal distance. For example, if a hydraulic system has an input piston with a reaction area of 1 square inch (645.2 square millimeters) and an output piston of 1 square inch (645.2 square millimeters), the output piston will move 1 inch (25.4 millimeters) if the input piston is moved 1 inch (25.4 millimeters). At the same time, if a force of 50 pounds (222.4 newtons) is applied to the input piston, the available force at the output piston will also be 50 pounds (222.4 newtons). Output force can be increased by increasing the size of the output piston. If, for instance, the output piston had a reaction area of 2 square inches, the available output force would be 50 pounds per square inch multiplied by the square inches of the piston, or 100 pounds of force. However, the distance traveled by the output piston would be reduced by 50 percent. The hydraulic system's efficiency depends to a great extent on the total absence of air or vapor in the system. If air or vapor is present in the system, the compressibility of these gases causes a loss in pressure, force, and motion.

Friction

The brake system depends on the principle of *friction* for operation. *Sliding friction* (kinetic friction) is the rubbing action of one

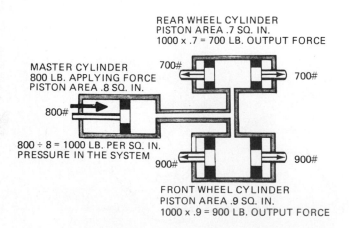

REAR WHEEL CYLINDER
PISTON AREA .7 SQ. IN.
1000 x .7 = 700 LB. OUTPUT FORCE

MASTER CYLINDER
800 LB. APPLYING FORCE
PISTON AREA .8 SQ. IN.

700#

700#

800#

800 ÷ 8 = 1000 LB. PER SQ. IN.
PRESSURE IN THE SYSTEM

900#

900#

FRONT WHEEL CYLINDER
PISTON AREA .9 SQ. IN.
1000 x .9 = 900 LB. OUTPUT FORCE

Figure 3-4. This schematic shows how a hydraulic system transmits force and motion. By varying output piston size, the available output force can be either increased or decreased as illustrated.

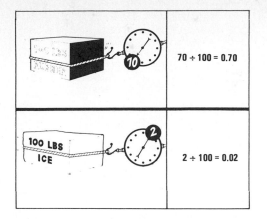

70 ÷ 100 = 0.70

100 LBS
ICE

2 ÷ 100 = 0.02

Figure 3-5. Coefficient of friction is calculated by dividing the weight of the object into the number of pounds of pull required to move the object. *(Courtesy of Ford Motor Co. of Canada Ltd.)*

object sliding on the surface of another, such as a brake disc on a brake pad or a brake drum on a brake lining. *Static friction* is the resistance to the sliding of one object on the surface of another. In a brake system, when the brakes are applied there is sliding friction between the shoe and drum or the disc and pad, as long as the wheels are still turning, and there is static friction between the tire and the road surface. Upon severe braking, the wheels become locked, at which time there is a sliding friction between the tire and the road and static friction between the shoe and drum or disc and pad.

The most effective braking takes place just before wheel lock-up occurs. The amount of *pressure* applied to the surfaces in contact determines the amount of friction present. More pressure means more friction; less pressure, less friction. If, for example, it takes a 60-pound pull to slide a 100-pound object over a dry, stationary surface, the resultant *coefficient of friction* would be 60/100 or 0.60. It would therefore take only 30 pounds of pull to slide 50 pounds of the same material over the same surface. Less weight, which has resulted in less pressure between the two surfaces, results in less friction.

Heat

Friction produces *heat*. More friction produces more heat. In the brake system, the energy (momentum or kinetic energy) of the moving vehicle is converted to heat energy by the friction in the brake system. This heat energy or heat must be dissipated by the brake drum and linings or rotor and pad. Repeated

severe braking results in excessive heat build-up in brake parts and causes *brake fade*. Brake fade is a condition that the driver recognizes as requiring excessive brake pedal pressure, which results in little or no braking. This condition is caused by the change in the coefficient of friction between lining and drum or disc and pad. Pads and linings become glazed, while drum and disc surfaces become hardened.

The ability of a brake system to dissipate heat depends on brake design. The size of the surface area of the friction elements, pads, discs, linings, and drums is a factor. Since disc brake friction surfaces are more exposed to atmosphere, they dissipate heat more effectively than drum brakes of similar size. Since about 60 percent (more on front-wheel-drive vehicles) of braking is done by the front brakes (due to forward weight transfer during brak-

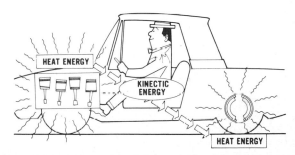

Figure 3-6. Energy conversion. The fuel's heat energy is converted to mechanical energy by the engine. The engine's mechanical energy is used to put the vehicle in motion (kinetic energy). The kinetic energy of the moving vehicle is converted to heat energy by the brake system to bring the vehicle to a stop. *(Courtesy of Ford Motor Co. of Canada Ltd.)*

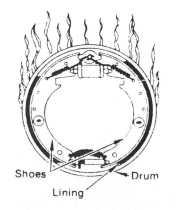

Shoes

Drum

Lining

Figure 3-7. Heat energy produced by the brakes must be dissipated to atmosphere. *(Courtesy of Ford Motor Co. of Canada Ltd.)*

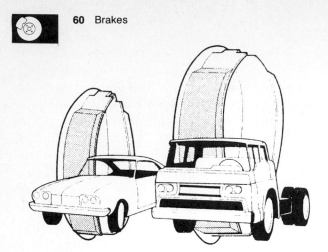

Figure 3-8. The larger or heavier the vehicle, the greater the friction area required to dissipate the heat generated by the brakes. *(Courtesy of Ford Motor Co. of Canada Ltd.)*

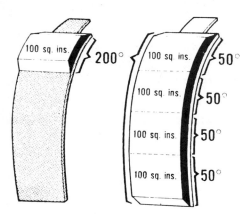

Figure 3-9. The larger the friction area in the brake system, the less heat is generated per square inch. *(Courtesy of Ford Motor Co. of Canada Ltd.)*

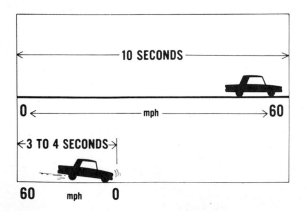

Figure 3-10. A vehicle that can accelerate from 0 to 60 mph in 60 seconds with a 100-horsepower engine is expected to come to a stop in as little as 3 to 4 seconds. This requires the brakes to dissipate approximately 1,500 horsepower of heat energy in 3 to 4 seconds. *(Courtesy of Ford Motor Co. of Canada Ltd.)*

ing), disc brakes are usually used in the front, and drum brakes, at the rear. The stopping distance of a vehicle depends on driver reac-

VEHICLE SPEED M.P.H.	DISTANCE TRAVELED DURING DRIVER'S REACTION TIME	DISTANCE TRAVELED AFTER APPLICATION OF BRAKES	TOTAL STOPPING DISTANCE
20	22 FT.	22 FT.	44 FT.
30	33 FT.	50 FT.	83 FT.
40	44 FT.	88 FT.	132 FT.
50	55 FT.	138 FT.	193 FT.
60	66 FT.	200 FT.	266 FT.

Figure 3-11. This table shows approximate stopping distance required to bring a vehicle to a stop at various speeds. Note the great increase in distance travelled as speed is increased.

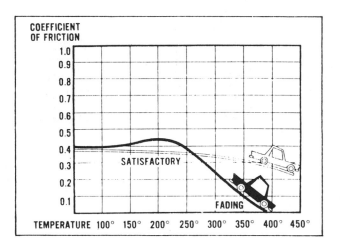

Figure 3-12. Severe braking can cause brakes to fade due to overheating. Overheating causes the characteristics of the friction surfaces to change so there is less friction. Little or no braking can result even with heavy pedal pressure. This is called *brake fade.*

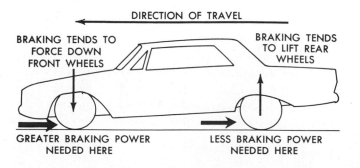

Figure 3-13. This shows why greater braking power is required at the front wheels of the car.

tion time, weight of the vehicle, and speed. If a vehicle's weight is doubled, the stopping distance required is doubled. If both weight and speed are doubled, approximately eight times the stopping distance is required.

PART 2 BRAKE HYDRAULIC SYSTEM

Master Cylinder (Front-Rear Split System)

When the driver pushes down on the brake pedal, the master cylinder pushrod pushes the primary piston into the master cylinder. This in turn forces the secondary piston deeper into the master cylinder. Fluid is forced out of

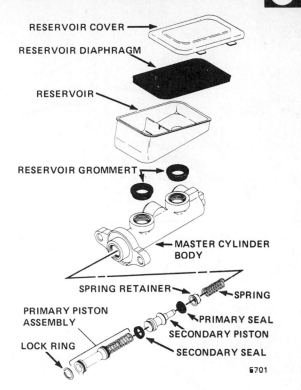

Figure 3-15. Exploded view of tandem master cylinder. *(Courtesy of General Motors Corporation)*

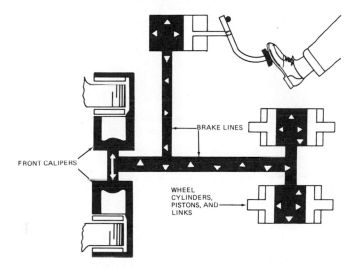

Figure 3-14. Hydraulic brake system operation with rear drum brakes and front disc brakes.

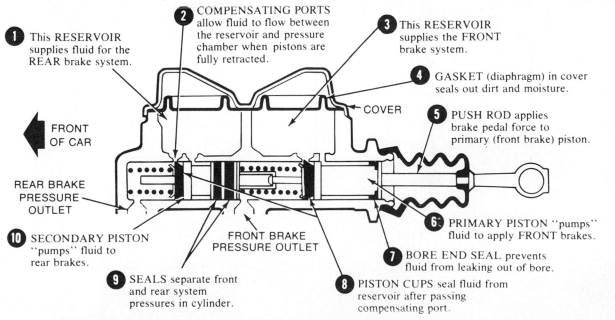

- Separate reservoirs, cylinders and pistons for front and rear systems.

1 This RESERVOIR supplies fluid for the REAR brake system.

2 COMPENSATING PORTS allow fluid to flow between the reservoir and pressure chamber when pistons are fully retracted.

3 This RESERVOIR supplies the FRONT brake system.

4 GASKET (diaphragm) in cover seals out dirt and moisture.

COVER

5 PUSH ROD applies brake pedal force to primary (front brake) piston.

FRONT OF CAR

REAR BRAKE PRESSURE OUTLET

10 SECONDARY PISTON "pumps" fluid to rear brakes.

FRONT BRAKE PRESSURE OUTLET

9 SEALS separate front and rear system pressures in cylinder.

6 PRIMARY PISTON "pumps" fluid to apply FRONT brakes.

7 BORE END SEAL prevents fluid from leaking out of bore.

8 PISTON CUPS seal fluid from reservoir after passing compensating port.

Figure 3-16. Cutaway view of tandem master cylinder with parts identified. *(Courtesy of Ford Motor Co. of Canada Ltd.)*

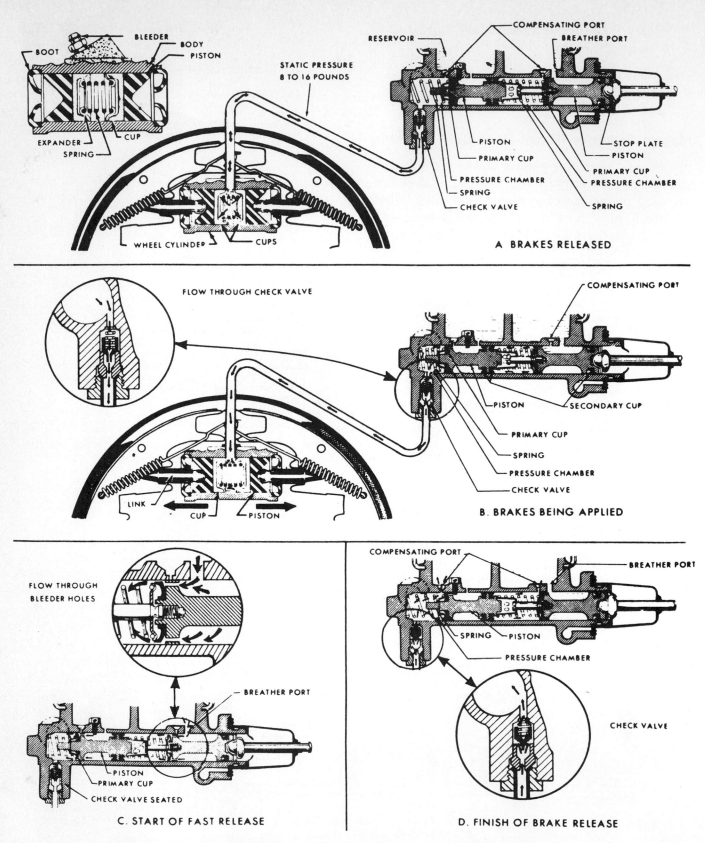

BOOT **BLEEDER** **BODY**

PISTON

EXPANDER **CUP**

SPRING

STATIC PRESSURE 8 TO 16 POUNDS

RESERVOIR **COMPENSATING PORT** **BREATHER PORT**

PISTON

PRIMARY CUP

PRESSURE CHAMBER

SPRING

CHECK VALVE

STOP PLATE

PISTON

PRIMARY CUP

PRESSURE CHAMBER

SPRING

WHEEL CYLINDER **CUPS**

A BRAKES RELEASED

FLOW THROUGH CHECK VALVE

COMPENSATING PORT

PISTON

SECONDARY CUP

PRIMARY CUP

SPRING

PRESSURE CHAMBER

CHECK VALVE

LINK

CUP **PISTON**

B. BRAKES BEING APPLIED

FLOW THROUGH BLEEDER HOLES

BREATHER PORT

PISTON

PRIMARY CUP

CHECK VALVE SEATED

C. START OF FAST RELEASE

COMPENSATING PORT

BREATHER PORT

SPRING **PISTON**

PRESSURE CHAMBER

CHECK VALVE

D. FINISH OF BRAKE RELEASE

Figure 3-17. Dual (tandem) master cylinder operation.

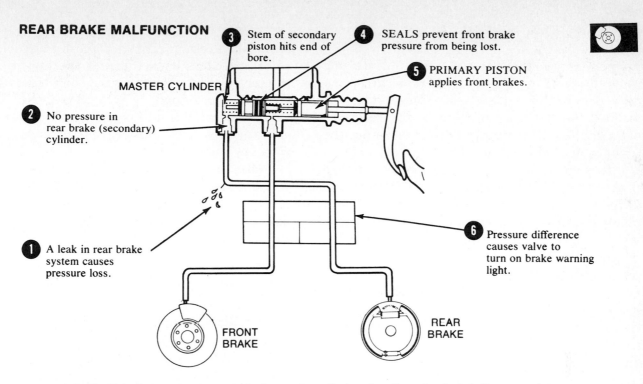

REAR BRAKE MALFUNCTION

3 Stem of secondary piston hits end of bore.

4 SEALS prevent front brake pressure from being lost.

5 PRIMARY PISTON applies front brakes.

MASTER CYLINDER

2 No pressure in rear brake (secondary) cylinder.

1 A leak in rear brake system causes pressure loss.

6 Pressure difference causes valve to turn on brake warning light.

FRONT BRAKE

REAR BRAKE

Figure 3-18. This shows what happens in the master cylinder when there is a leak in the secondary hydraulic system, such as in the rear brake lines or rear-wheel cylinders. *(Courtesy of Ford Motor Co. of Canada Ltd.)*

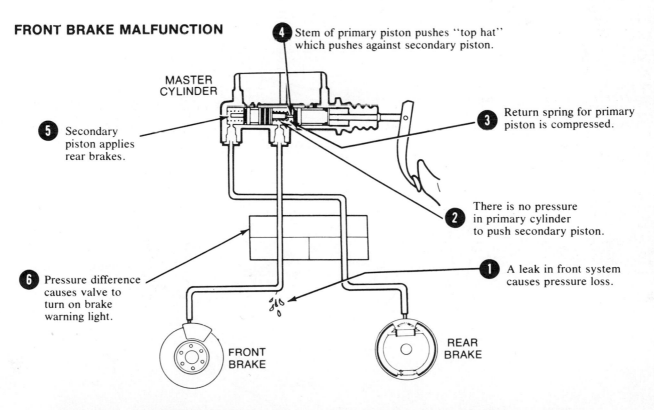

FRONT BRAKE MALFUNCTION

4 Stem of primary piston pushes "top hat" which pushes against secondary piston.

MASTER CYLINDER

5 Secondary piston applies rear brakes.

3 Return spring for primary piston is compressed.

2 There is no pressure in primary cylinder to push secondary piston.

6 Pressure difference causes valve to turn on brake warning light.

1 A leak in front system causes pressure loss.

FRONT BRAKE

REAR BRAKE

Figure 3-19. This illustration shows how the dual master cylinder operates when there is a hydraulic leak in the primary system, such as in the lines to the front brakes or in the front-wheel cylinders or calipers. *(Courtesy of Ford Motor Co. of Canada Ltd.)*

the primary system through steel and flexible lines to the front brakes and from the secondary system through separate lines to the rear brakes. As the level of fluid in the reservoir rises and falls, the diaphragm flexes, since the cover is vented to atmosphere.

In normal operation, the secondary piston is actuated hydraulically. Should a fluid leak develop in the primary system, the primary piston pushes against the secondary piston since hydraulic pressure in the primary piston is lost. Should a leak develop in the secondary system, the secondary piston "bottoms out" in the master cylinder, and the primary system still operates normally. In either case the brake pedal will be somewhat lower than normal, and the brake warning light will go on when the pedal is pushed down.

When the brake pedal is released, the brake shoe return springs force fluid from the wheel cylinders back through the lines to the master cylinder. When pressure in the system drops to approximately 8 to 18 psi, the check valve closes and fluid flow stops. This residual or static pressure helps seal piston cups in the wheel cylinders.

Disc brake systems do not have check valves since disc brake caliper pistons have a different seal. Some drum brake systems use mechanical wheel cylinder piston cup expanders and thereby eliminate the need for static pressure in the system.

Diagonally Split Hydraulic System

This system is used on vehicles where the weight distribution is such that the greater proportion of weight acts on the front wheels. This is the case with front-engine, front-wheel-

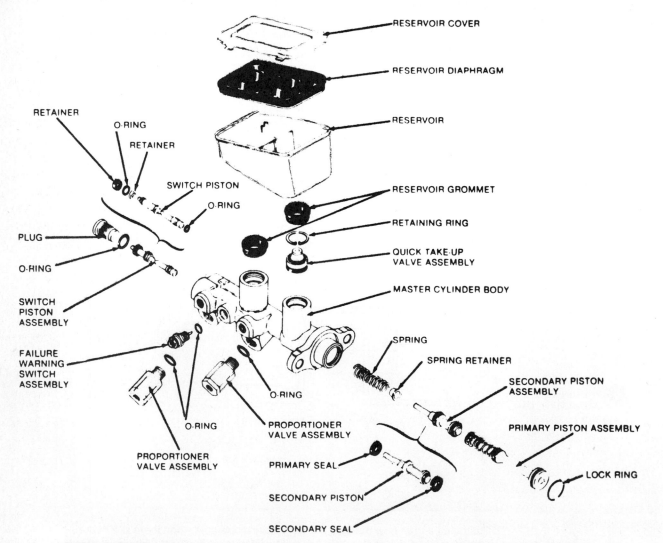

Figure 3-20. Exploded view of quick take-up master cylinder used on diagonally split hydraulic system on front-wheel-drive vehicle. *(Courtesy of General Motors Corporation)*

drive vehicles and vehicles with a high degree of weight transfer from the rear to the front during braking.

With a diagonally split system, the left front wheel and the right rear wheel would be linked hydraulically to one half of the master cylinder while the right front wheel and left rear wheel would be linked hydraulically to the other half of the master cylinder. Each of the two halves of the diagonally split hydraulic system operates independently of the other half, just as in the front-rear dual hydraulic system, and provides a high degree of safety in case of partial brake failure.

In this design, when one half of the system fails, the other half would still provide braking at one front wheel and one rear wheel. This creates a brake imbalance, which would result in a pull to the side of the vehicle where the front brake is still functioning. However, to offset this imbalance, a negative scrub radius is designed into the front suspension system. This negative scrub radius also compensates for any difference in tire-to-road adhesion during braking caused by differences in road surface conditions between front wheels, i.e., ice, gravel, etc.

Quick Take-Up Master Cylinder

This master cylinder is used in a diagonal split system. It incorporates the functions of the standard dual master cylinder plus a warning light switch and proportioners. It incorporates a quick take-up feature that provides a large volume of fluid to the wheel brakes at low pressure with the initial brake application. The low-pressure fluid quickly provides the fluid displacement requirements of the system created by the seals retracting the pistons into the front calipers and retraction of rear drum brake shoes. The quick take-up master cylinder operates as follows.

1. With the initial brake application, more fluid is displaced in the primary piston low-pressure chamber than in the high-pressure chamber since the low-pressure chamber has a larger diameter. The additional fluid is forced around the outside diameter of the primary piston lip seal, into the high-pressure chamber, and on to the wheel brake units. Since equal pressure and displacement must be maintained in both primary and secondary systems, the primary piston moves a shorter distance than the secondary piston to compensate for the larger volume of fluid moved from the low-pressure area of the primary piston to the high-pressure area.

2. As the low-pressure displacement requirements are met, pressure will increase in the primary piston low-pressure chamber until the spring-loaded ball check valve in the quick take-up valve opens. This allows fluid to flow into the reservoir.

3. After the quick take-up phase of the cycle is completed, the pistons function in the same manner as in a conventional dual master cylinder.

4. With the release of the brakes, the master cylinder springs will return the master cylinder pistons faster than fluid can flow back through the systems. This would tend to create a vacuum on both the low-pressure and high-pressure chambers of the pistons if proper compensation were not provided.

5. The primary piston is compensated by fluid flowing from the reservoir through the small diameter holes of the quick take-up valve around the outside diameter of the quick take-up lip seal through the bypass hole and compensating port and into the low- and high-pressure chambers of the primary piston. The secondary piston is compensated by fluid flowing from the reservoir through the bypass hole and compensating port into the high- and low-pressure areas.

6. Expansion and contraction of brake fluid are handled by fluid passing directly from the master cylinder bore through the bypass hole and compensating port to the reservoir in a conventional dual bore master cylinder. The secondary piston in the quick take-up master cylinder functions in this same manner. However, the primary piston must work through the quick take-up valve; thus a bypass groove is used to account for the fluid flow from or to the primary piston chambers.

Wheel Cylinders

Wheel cylinders are used in drum brakes to force the brake shoes against the drum. Front-wheel cylinders are usually larger in

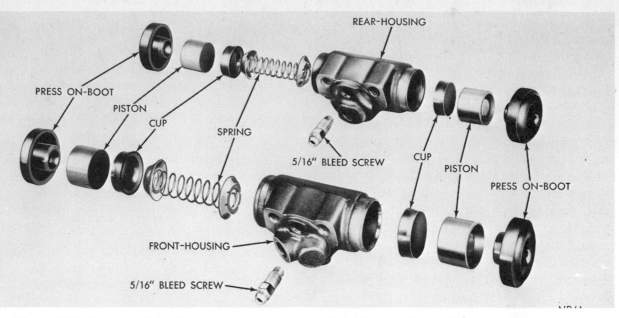

PRESS ON-BOOT
PISTON
CUP
SPRING
5/16" BLEED SCREW
REAR-HOUSING
CUP
PISTON
PRESS ON-BOOT
FRONT-HOUSING
5/16" BLEED SCREW

Figure 3-21. Exploded view of rear drum brake wheel cylinder (top) and front drum brake wheel cylinder (bottom). Note the difference in piston diameters resulting in more output at the front brakes, which do most of the braking. *(Courtesy of Chrysler Corporation)*

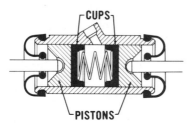

CUPS
PISTONS

Figure 3-22. Cutaway view of assembled wheel cylinder. Note piston cups must face spring to seal hydraulic fluid pressure properly. *(Courtesy of Ford Motor Co. of Canada Ltd.)*

diameter than rear-wheel cylinders since most of the braking is done at the front wheels. Each wheel cylinder is fitted with a spring, cups, pistons, pushrods, and dust boots. A fitting is provided to which the brake line is attached, and a bleeder is used to bleed air from the system. When brakes are applied, the pistons are forced outward to push the shoes against the drum. On brake release, the brake shoe return springs force the fluid out of the wheel cylinder by pushing the pistons into the cylinder. Wheel cylinders are mounted to the backing plate.

Wheel Cylinder Design and Operation

Wheel cylinder designs include dual-piston, single-bore cylinders; dual-piston, step-bore cylinders; and single-piston cylinders.

The dual-piston, single-bore wheel cylinder is used with the dual-servo drum brake design. The step-bore, dual-piston wheel cylinder may be used with some nonservo drum brake designs. The larger diameter piston acts on the secondary or trailing shoe to increase apply force since self-energization has a tendency to resist hydraulic apply force on this design.

The single-piston wheel cylinder is used on some dual-servo design brakes and on some nonservo double-leading shoe drum brakes.

The spring in the wheel cylinder holds the rubber cup against the piston and the piston against the push rod or brake shoe, depending on design. The sealing edge of the rubber cup is flared (larger in diameter than the wheel cylinder bore) to help prevent fluid leakage and entry of air.

Usually, residual hydraulic pressure in the system or metal disc cup expanders or both increase the sealing pressure of the cup against the cylinder bore. When the brakes are applied, increased hydraulic pressure increases the sealing pressure of the cup lip against the cylinder bore to prevent any escape of fluid. Obviously, if the rubber cup were to be installed the wrong way, this action would not take place and fluid would be forced past the cup out of the hydraulic system.

The cylinder bore area between the rubber cups is subject to corrosion and rust especially if brake fluid becomes contaminated. As the brake linings wear, the distance between the cups increases and the possibility for corrosion in a larger area of the cylinder bore is obvious. Should new brake shoes or linings be installed without servicing the wheel cylinders, the wheel cylinder piston cups would be pushed back and would be forced to operate on the rusted or corroded area of the cylinder bore. This would soon result in cup damage and hydraulic fluid leakage with resultant brake failure.

Wheel cylinders are equipped with dust boots or seals. In some designs the boot fits around the outside of the wheel cylinder whereas on others the boot fits on the inside of the end of the wheel cylinder bore. The dust boot prevents dirt and moisture from entering the wheel cylinder bore area, where the piston must travel back and forth.

If the operation of the wheel cylinders and the brake hydraulic system are properly understood the reasons for good service procedures are also understood and can be better explained to the customer or vehicle owner.

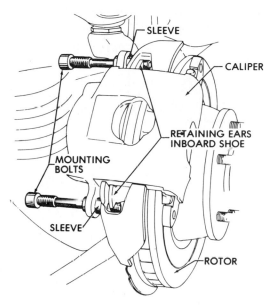

Figure 3-23. Caliper, guide pins, and bushings shown positioned over the rotor (disc). *(Courtesy of General Motors Corporation)*

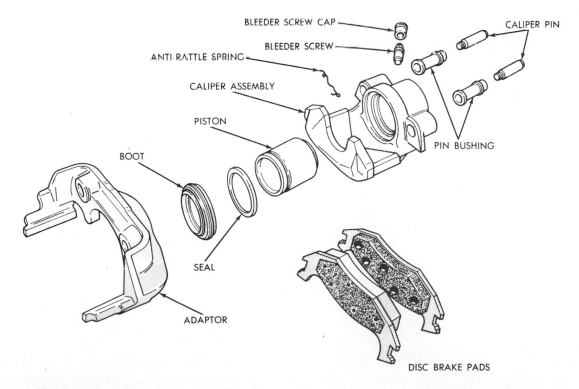

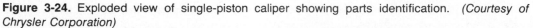

Figure 3-24. Exploded view of single-piston caliper showing parts identification. *(Courtesy of Chrysler Corporation)*

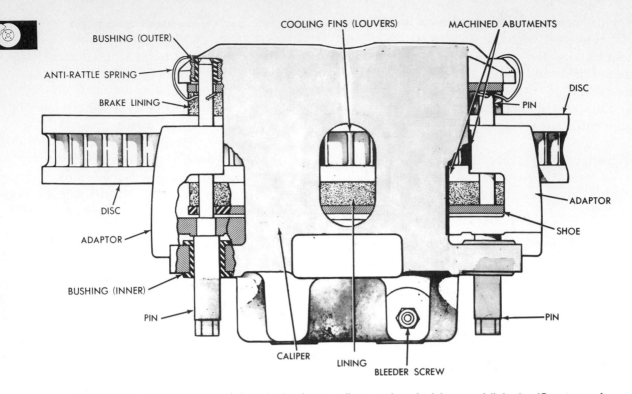

Figure 3-25. Another view of the single-piston caliper and pads (shoe and lining). *(Courtesy of Chrysler Corporation)*

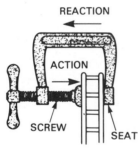

Figure 3-26. Single-piston floating caliper acts like a C-clamp exerting equal pressure on both sides of disc when brakes are applied.

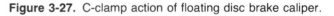

Figure 3-27. C-clamp action of floating disc brake caliper.

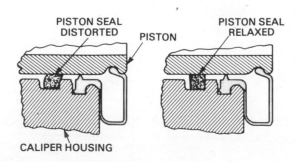

Figure 3-28. Piston seal distorts when brakes are applied and resumes normal relaxed position, retracting the piston, when brakes are released. This provides proper pad-to-rotor clearance.

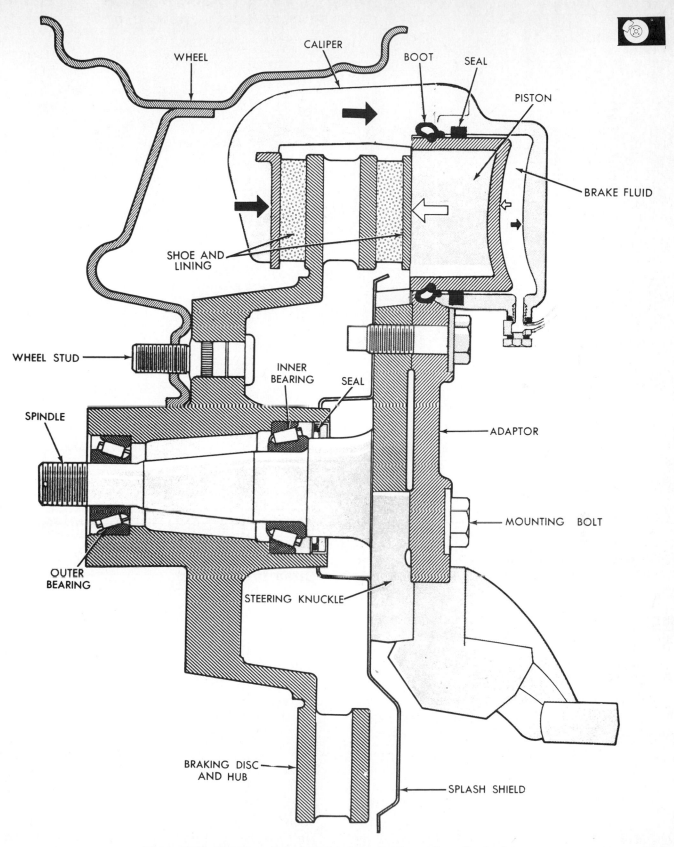

Figure 3-29. Cutaway view of single-piston floating caliper disc brake.

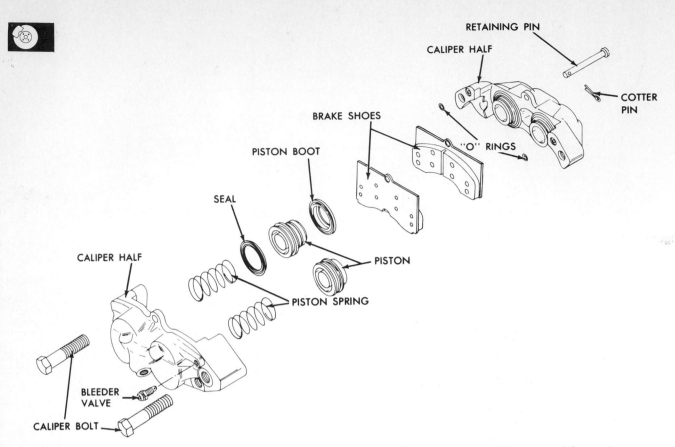

Figure 3-30. Four piston, fixed caliper as used on some older disc brakes. *(Courtesy of General Motors Corporation)*

Calipers

The disc brake caliper is attached to the disc brake adapter (anchor plate). The caliper consists of a housing and cylinder with a piston and seal in the cylinder. The disc brake pads are also mounted in the calipers. The floating caliper usually has only one piston on the inboard side and acts on the disc or rotor like a C-clamp tightening on it. The caliper can move slightly in or out, thereby exerting equal pressure to both sides of the disc. The caliper either floats on pins and rubber bushings or slides on metal guides. When the brakes are applied, the fluid forces the piston out, pushing the pad against the disc. The reaction on the caliper causes it to move the other pad inward slightly, applying equal pressure to the other side of the disc. On brake release the piston seal resumes its normal relaxed shape, thus pulling the piston back slightly. The piston maintains proper pad-to-disc clearance by moving out as the pads wear, thereby providing the self-adjusting feature. A connection for the brake line and a bleeder are also provided.

PART 3   DISC AND DRUM BRAKES

Drum Brake Units

On the typical dual-servo drum brakes the entire braking assembly is mounted on a backing plate. This backing plate is bolted to the axle housing at the rear and to the steering knuckle at the front. The shoe anchor is attached to the backing plate and must absorb all the braking torque developed as the shoes are applied to the drums.

The shoes are mounted to the backing plate by hold-down pins and springs and are able to slide back and forth slightly on the backing plate ledges. A star-wheel adjuster is positioned between the shoes at the bottom, and return springs hold the shoes against the anchor at the top. A wheel cylinder is mounted to the backing plate just below the anchor to push the shoes against the drum during braking. The entire assembly is enclosed with a brake drum that is attached to the wheel hub

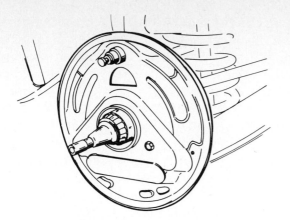

Figure 3-31. Backing plate mounted on front-wheel steering knuckle.

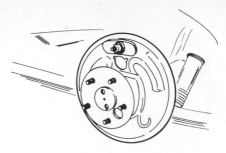

Figure 3-32. At the rear wheels, backing plates are mounted on flanges at the ends of the axle housing.

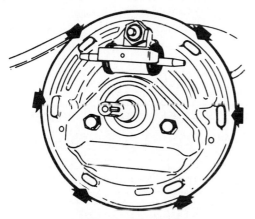

DRESS ALL SUPPORT PADS

Figure 3-33. Backing plate ledges or platforms keep the shoes in proper alignment with the backing plate and brake drum.

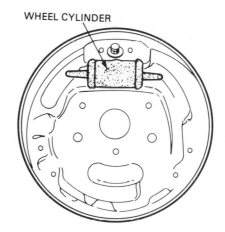

WHEEL CYLINDER

Figure 3-34. A hydraulic wheel cylinder is mounted on the backing plate. When the brake pedal is pushed down, hydraulic pressure inside the wheel cylinder forces the two pistons to move outward slightly, pushing against the push-rods and forcing the shoes against the drum.

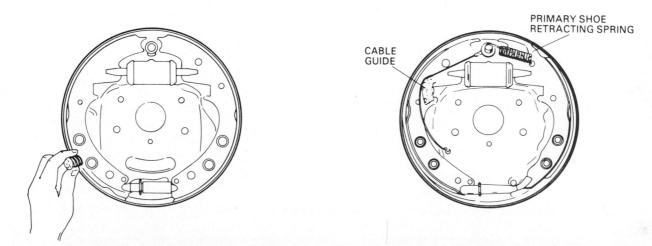

CABLE GUIDE

PRIMARY SHOE RETRACTING SPRING

Figure 3-35. The shoe assembly is positioned on the backing plate and secured with hold-down pins, springs, and cups. The assembly is able to slide back and forth slightly on the backing plate. Retracting springs hold the shoes against the anchor.

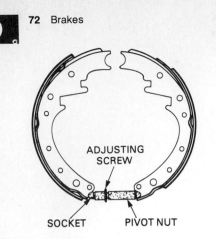

Figure 3-36. Star-wheel adjuster can be lengthened or shortened by turning the adjusting screw to provide proper shoe-to-drum clearance. Adjusters for the left side are not interchangeable with adjusters for the right side of the vehicle.

or axle flange. Two types of self-adjusters are in use: a cable-operated type and a lever-operated type. When the brakes are applied while the car is backing up, the rear brake shoe is pulled away from the anchor. This ac-

tuates the cable or levers, which in turn move the star-wheel adjuster. This happens when enough lining wear has taken place to provide enough shoe movement to move the star wheel adjuster one notch.

The front brake shoe is known as the *primary shoe* and the rear shoe on the same wheel is known as the *secondary shoe*. The primary shoe usually has a shorter lining, often of a different material than the secondary lining, because the secondary shoe must do most of the braking. Since the shoe-and-lining assembly are floating and self-centering in the drum, the primary shoe pushes against the secondary shoe through the adjuster. When the brakes are applied, the primary shoe tends to bite deeper into the drum and wants to rotate with the drum, thereby pushing the secondary shoe much tighter against the drum. The tendency for the shoe to bite deeper into the drum is called self-energizing. The primary shoe pushing against the secondary shoe is called servo-action.

Brake linings either are bonded to the shoe under heat and pressure or are riveted

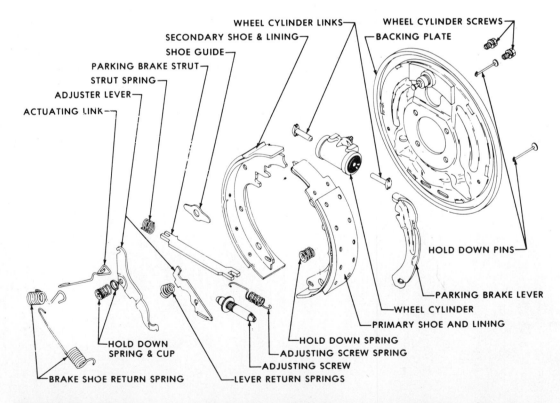

Figure 3-37. Different types of self-adjusting mechanisms and their application. Operation is similar on all. As the vehicle moves in reverse and the brakes are applied, the secondary shoe moves away from the anchor. This actuates the self-adjusting linkage. When lining wear allows sufficient linkage movement, the adjuster will move one notch. This adjusts the shoes closer to the brake drum. *(Courtesy of General Motors Corporation and Chrysler Corporation)* [Figure 3-37 continued on page 73.]

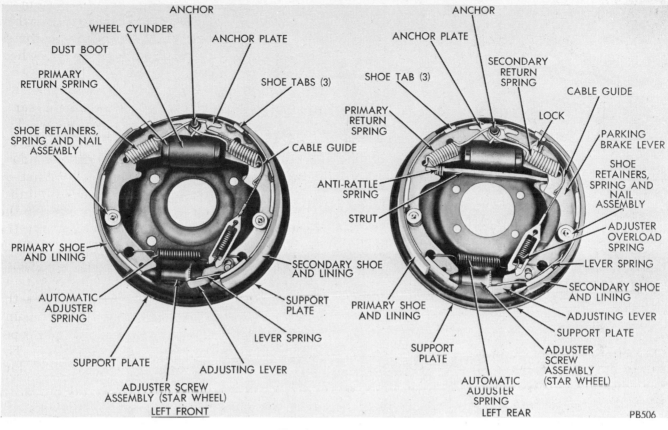

Figure 3-37. *(Continued)*

to the shoe. Most standard brake linings are organic and contain a large amount of asbestos with a number of additives for bonding and stabilizing. Heavy-duty linings are usually metallic. Under severe braking, temperatures may reach 500° F. (260° C.) or more. Metallic brake linings are better able to withstand higher temperatures. Brake linings can crack and separate from the shoe under severe braking conditions, such as descending a mountain. The brake shoe itself is made of metal and is very rigid since it must maintain its shape under high pressures.

Brake drums are cast iron with a steel disc or web. Many drums are finned for better cooling and some have spring-type vibration

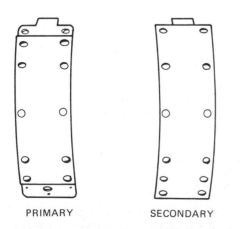

Figure 3-38. Primary shoe with shorter lining is mounted at front position while longer lining secondary shoe is mounted at rear position of each wheel.

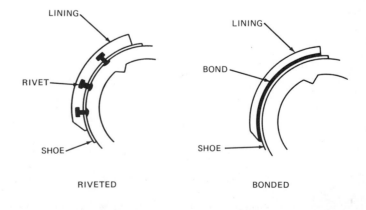

Figure 3-39. Methods of attaching brake linings to brake shoes. Rivet holes in riveted linings are countersunk about two-thirds of lining thickness. Tubular rivets are used. Bonded linings are glued to the shoe, clamped in place, then cured in an oven.

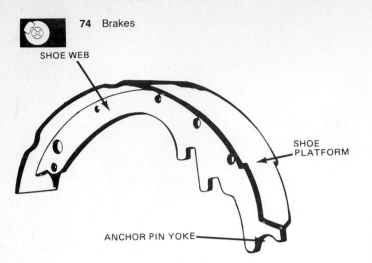

SHOE WEB

SHOE PLATFORM

ANCHOR PIN YOKE

Figure 3-40. Typical brake shoe before lining has been attached.

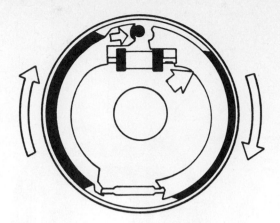

Figure 3-42. Self-energizing action of primary shoe pushes against secondary shoe through the adjuster at the bottom (servoaction). As a result, the secondary shoe does more of the braking. Secondary shoe rotation is prevented by the anchor at the top.

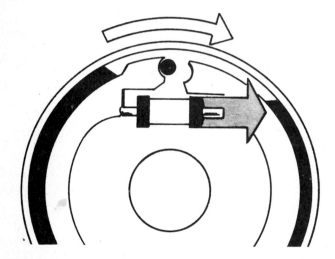

Figure 3-41. Wheel cylinder forces primary shoe against drum. Drum rotation pulls primary shoe away from anchor. Drum rotation also causes primary shoe to bite deeper into drum (self-energizing).

Figure 3-43. Heavier cars often use finned brake drums to aid in cooling the brakes.

dampers. Drums can become out-of-round, distorted, bell-mouthed, or barrel-shaped due to wear and abuse.

The secondary shoe at the rear wheels has a parking lever attached to it that is actuated by the parking brake cable. A parking brake link or strut is positioned between this lever and the primary shoe. The link is also equipped with an anti-rattle spring.

Drum Brake Design

Several different drum brake designs have been used. These are:

1. The dual-servo brake

2. The uniservo brake

3. The nonservo brake

Some minor variations exist in each of these designs such as wheel cylinder position, fixed or floating wheel cylinder, anchor mounting (on backing plate or direct to plate suspension component), anchor location on backing plate, type of adjuster, and type of shoe mounting.

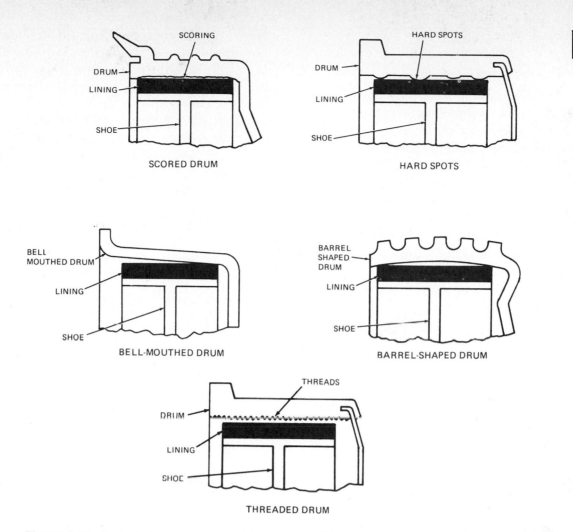

Figure 3-44. Scored drums are caused by worn-out linings or dirt. Hard spots are caused by overheating. Bell-mouthed and barrel-shaped drums result from shoe misalignment and worn-out drums. A threaded drum results from incorrect machining.

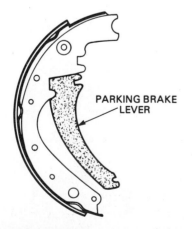

Figure 3-45. Parking brake lever attached to the secondary shoe. Lever pivots at its attaching pin and is held in place by a horseshoe-type retainer.

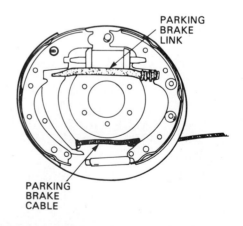

Figure 3-46. Parking brake cable is connected to the parking brake lever at the bottom. As the lever is pulled forward by the cable, it causes the link to push the primary shoe against the drum. Reaction at the lever pivot pushes the secondary shoe against the drum. When the parking brake is released, a spring on the cable pushes the lever back to its release position.

Dual-Servo Design

The dual-servo drum brake design uses one "dual"-piston wheel cylinder and two brake shoes, a primary shoe in the front position and a secondary shoe at the rear position. In this design the self-energizing action of the primary shoe pushes against the secondary shoe through the adjuster to increase braking apply force. The secondary shoe attempts to rotate until it contacts the anchor where braking torque is transmitted to the backing plate and the suspension system.

Uniservo Design

The uniservo drum brake design is the same as the dual-servo design with one exception. It uses a single-piston wheel cylinder instead of a dual-piston wheel cylinder. The wheel cylinder is mounted so that brake apply force from the wheel cylinder acts on the primary shoe only. Self-energization and servo action increase braking force in the same manner as in the dual-servo design.

Nonservo Design

The nonservo drum brake design has the brake shoes operating entirely independently from each other. There is no servo action. Depending on design, there may or may not be self-energization.

On the leading-trailing shoe design, a dual-piston wheel cylinder is used to apply both shoes independently since each shoe is separately anchored. This arrangement does not provide any servo action. Self-energization takes place on the leading shoe only when brakes are applied during forward motion and on the trailing shoe only when brakes are applied during reverse vehicle movement. Braking torque is transmitted to the anchor from one shoe and to the wheel cylinder from the other.

On the two leading shoe designs each shoe is actuated by a separate single piston-wheel cylinder. The wheel cylinders are mounted to provide self-energization of both shoes upon brake application during forward movement of the vehicle. There is no self-energization when brakes are applied while the vehicle is moving in a reverse direction. Since each shoe

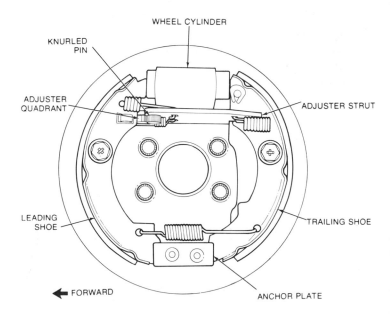

Figure 3-48. Nonservo rear drum brake unit with ratchet type of self-adjuster used by one manufacturer on a front-wheel-drive vehicle. *(Courtesy of Ford Motor Co. of Canada Ltd.)*

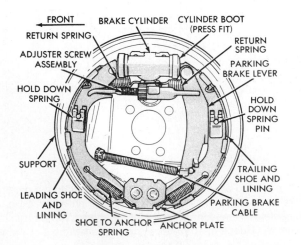

Figure 3-47. Nonservo rear drum brake unit with manual brake shoe adjuster used by one manufacturer on a front-wheel-drive vehicle. *(Courtesy of Chrysler Corporation)*

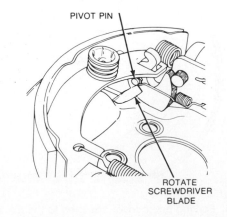

Figure 3-49. Backing off the brake adjustment with a screwdriver on the type of brake shown in Figure 3-48. *(Courtesy of Ford Motor Co. of Canada Ltd.)*

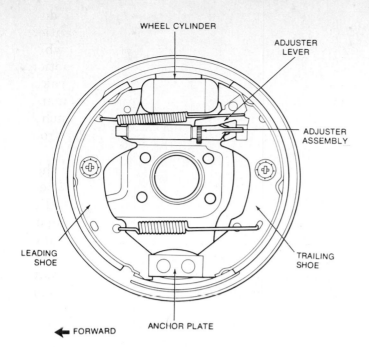

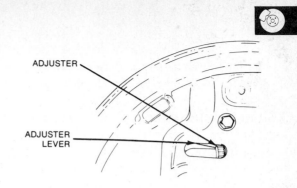

Figure 3-51. Access hole through backing plate allows backing off brake adjustment to facilitate drum removal on brake unit shown in Figure 3-50. *(Courtesy of Ford Motor Co. of Canada Ltd.)*

Figure 3-50. Nonservo rear drum brake unit used on front-wheel-drive vehicle by one manufacturer. This unit has a ratchet type of self-adjuster. *(Courtesy of Ford Motor Co. of Canada Ltd.)*

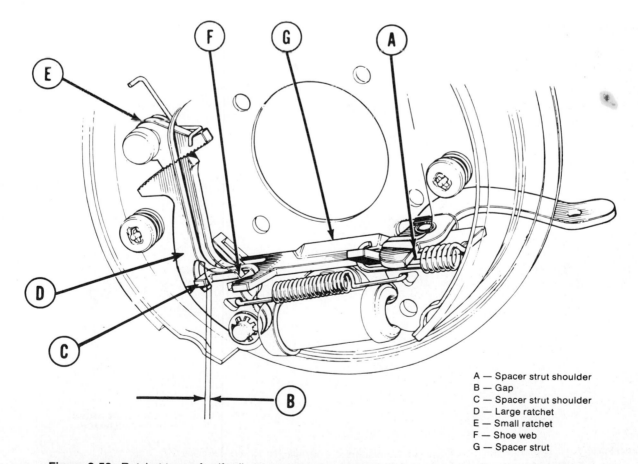

A — Spacer strut shoulder
B — Gap
C — Spacer strut shoulder
D — Large ratchet
E — Small ratchet
F — Shoe web
G — Spacer strut

Figure 3-52. Ratchet type of self-adjuster used on front-wheel-drive rear drum brake by one vehicle manufacturer. *(Courtesy of Ford Motor Co. of Canada Ltd.)*

is separately anchored, there is no servo action. Braking torque is transmitted to the two anchors during forward motion braking.

Drum Brake Adjustment

Shoe-to-drum clearance adjustment is provided in some designs by a star wheel screw-type adjuster between the two shoes and may be adjusted automatically, manually, or both. A ratchet cam and strut arrangement is also used in some designs for both manual and automatic adjustment. A direct acting cam or manually adjusted eccentric is also used on some vehicles. Periodic adjustment of lining-to-drum clearance is necessary to maintain pedal reserve. If the lining-to-drum clearance is excessive, the brake pedal may bottom out before the linings contact the brake drum. This condition is dangerous.

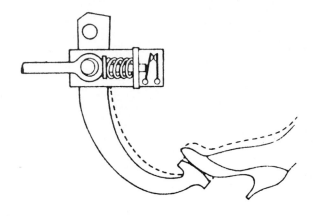

Figure 3-53. Pedal-actuated brake stop light switch completes electrical circuit to rear brake lights when pedal is depressed.

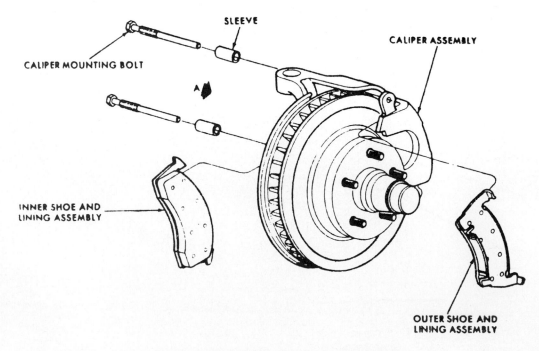

Figure 3-54. Floating caliper disc brake components.

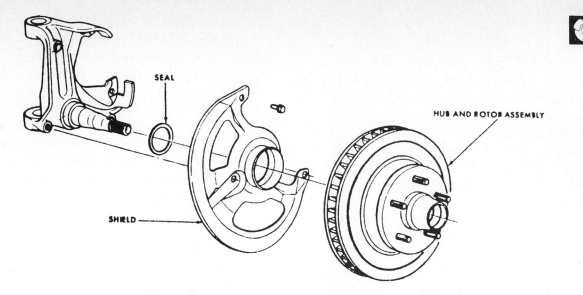

Figure 3-55. Steering knuckle, splash shield, and finned rotor for disc brake.

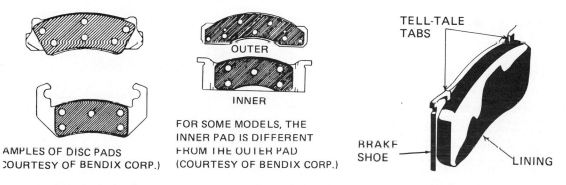

Figure 3-56. Various types of disc brake pads. When linings are worn down to telltale tabs, they cause a noise to be heard, reminding the driver that the brakes need attention. Friction materials used on pads are organic compounds, combination organic and metallic compounds, and metallic compounds.

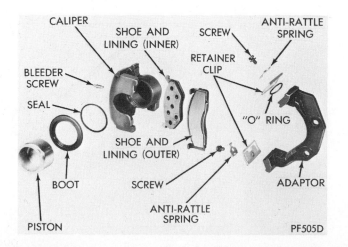

Figure 3-57. Disc brake caliper, exploded view. Antirattle clips for brake pads are also shown. *(Courtesy of Chrysler Corporation)*

Disc Brake Units

The disc brake unit consists of caliper, disc or rotor, pads, and splash shield. The caliper, described earlier in this chapter, is the unit that applies the brake pads against the disc. Earlier disc brakes used multipiston fixed calipers. Current disc brakes are mostly of the sliding or floating caliper single-piston type.

The most common application of disc brakes is on the front wheels, with drum brakes provided for the rear wheels. Some cars use four-wheel disc brakes. Calipers are bolted to a support bracket, which in turn is bolted to the steering knuckle. When braking during forward motion, braking torque attempts to push the upper end of the knuckle forward. The steering knuckle therefore must absorb all the braking torque developed at the front wheels.

Discs or rotors are either the solid or finned type. The finned type has better cooling and therefore is usually used on heavier cars. Disc brake surfaces, like drum brake surfaces, are highly machined and must remain smooth and parallel for effective braking. The disc is made of cast iron, which has proved to be the best friction material for both discs and drums. The minimum thickness for discs is often stamped in the disc. Brake drums also have maximum allowable diameters stamped on them. Disc brake pads are either bonded or riveted to a metal shoe. Pads are usually backed with a stick-on anti-rattle material on the metal shoe side and are often mounted with anti-rattle clips. Disc brakes are not self-energizing and therefore have large pistons

Figure 3-58. Using a screwdriver to adjust the parking brake shoe to drum clearance on the rear wheel of one type of four-wheel disc brake. See internal parts of this unit in Figure 3-59. *(Courtesy of General Motors Corporation)*

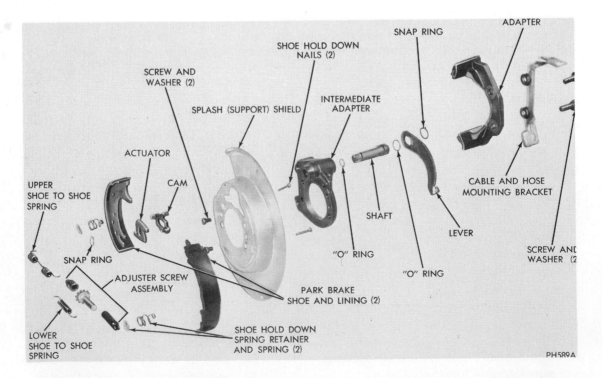

Figure 3-59. Internal expanding shoe type of parking brake used with some four-wheel disc brakes. The rear disc incorporates a small brake drum for the parking brake. *(Courtesy of Chrysler Corporation)*

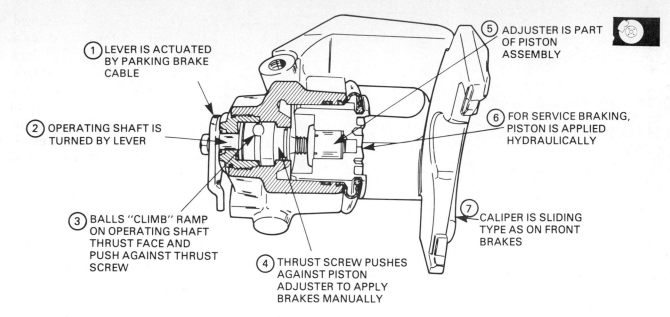

(1) LEVER IS ACTUATED BY PARKING BRAKE CABLE

(2) OPERATING SHAFT IS TURNED BY LEVER

(3) BALLS "CLIMB" RAMP ON OPERATING SHAFT THRUST FACE AND PUSH AGAINST THRUST SCREW

(4) THRUST SCREW PUSHES AGAINST PISTON ADJUSTER TO APPLY BRAKES MANUALLY

(5) ADJUSTER IS PART OF PISTON ASSEMBLY

(6) FOR SERVICE BRAKING, PISTON IS APPLIED HYDRAULICALLY

(7) CALIPER IS SLIDING TYPE AS ON FRONT BRAKES

Figure 3-60. Ford four-wheel disc brake parking brake operation.

to produce the necessary force for high apply pressures. This generates more heat than do drum brakes, but disc brakes are better able to dissipate the heat produced. The result is less *brake fade* than with drum brakes. Brake fade occurs when friction surfaces become so hot that their coefficient of friction drops so low that even the application of severe pedal pressure can result in little actual braking.

Four-Wheel Disc Brakes

Disc brakes are used on some vehicles on all four wheels because of the advantage disc brakes have over drum brakes. Rear-wheel disc brakes are similar in design to front-wheel disc brakes and they operate in essentially the same manner. The same type of components are used---calipers, pads, rotors---and the hydraulic system is the same. The major difference is that a parking brake is required at the rear wheels.

Some manufacturers use an internal expanding shoe and drum type of parking brake. The usual type of parking brake control mechanism is used to operate the parking brake. This unit consists of a backing plate on which two shoes are mounted in a similar manner as the duo-servo drum brake arrangement and a brake drum that is an integral part of the brake disc. A screw-type adjuster is used to provide proper shoe-to-drum clearance.

Another type of rear-wheel disc brake uses the disc brake as the parking brake. This is

accomplished by using a modified caliper assembly that allows mechanical application of the pads against the rotor. A lever-operated ramp- and-ball arrangement forces the caliper piston to move to apply the brakes manually (see Figure 3-60).

PART 4   BRAKE LINES, SWITCHES, VALVES AND FLUID

Brake Lines

Seamless steel brake lines carry the brake fluid from the master cylinder to a brake warning light switch or to a combination valve. From there, additional steel lines carry the fluid to the front wheel openings and to the rear of the frame or body. At the rear a high-pressure flexible line connects the steel line to other steel lines in the rear axle, which lead to the rear-wheel cylinders. High-pressure flexible lines are also used at each front-wheel cylinder or caliper. Only recommended steel and flex lines should be used for replacement.

Two types of steel line flared ends are used. One type is the double lap flare and the other type is the ISO flare. Fittings for these two types are also of different design and should never be mixed or interchanged. Never use copper or aluminum lines; they can fail due to the high pressure developed in the brake system.

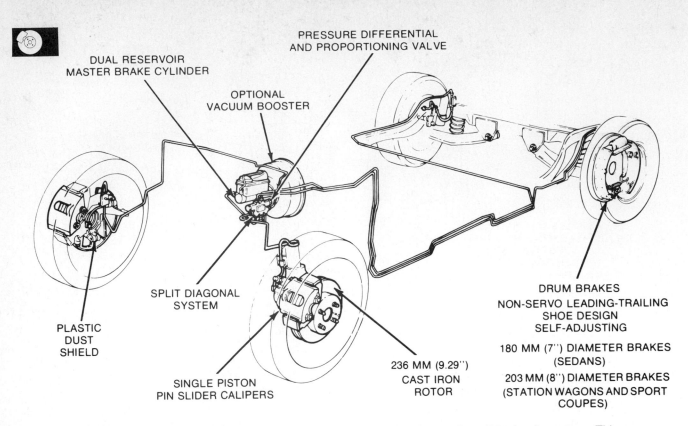

DUAL RESERVOIR
MASTER BRAKE CYLINDER

PRESSURE DIFFERENTIAL
AND PROPORTIONING VALVE

OPTIONAL
VACUUM BOOSTER

SPLIT DIAGONAL
SYSTEM

PLASTIC
DUST
SHIELD

SINGLE PISTON
PIN SLIDER CALIPERS

236 MM (9.29'')
CAST IRON
ROTOR

DRUM BRAKES
NON-SERVO LEADING-TRAILING
SHOE DESIGN
SELF-ADJUSTING

180 MM (7'') DIAMETER BRAKES
(SEDANS)

203 MM (8'') DIAMETER BRAKES
(STATION WAGONS AND SPORT
COUPES)

Figure 3-61. Brake lines on a front-wheel-drive vehicle with a diagonally split hydraulic system. This system requires two brake lines to the rear of the vehicle. *(Courtesy of Ford Motor Co. of Canada Ltd.)*

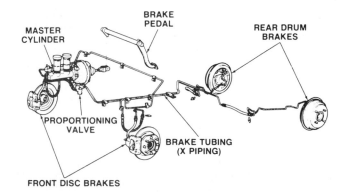

MASTER
CYLINDER

BRAKE
PEDAL

REAR DRUM
BRAKES

PROPORTIONING
VALVE

BRAKE TUBING
(X PIPING)

FRONT DISC BRAKES

Figure 3-62. Service brake, power disc, and drum type. *(Courtesy of Ford Motor Co. of Canada Ltd.)*

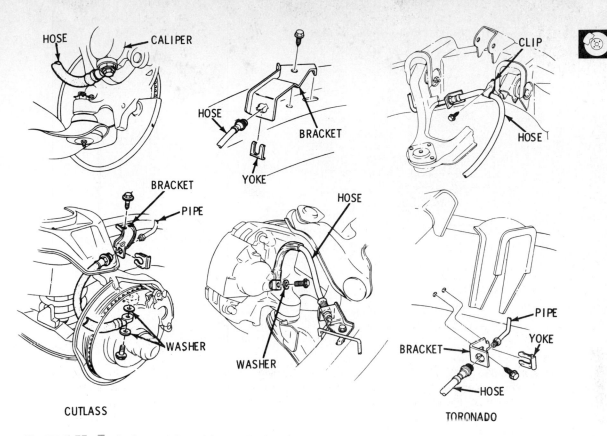

Figure 3-63. Typical examples of front-wheel brake line mounting methods. *(Courtesy of General Motors Corporation)*

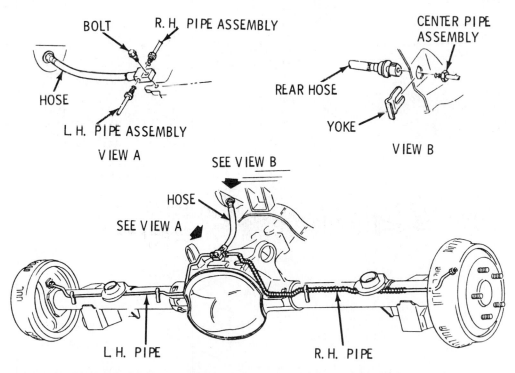

Figure 3-64. Typical brake line routing on a rear axle of a front-rear split hydraulic brake system. *(Courtesy of General Motors Corporation)*

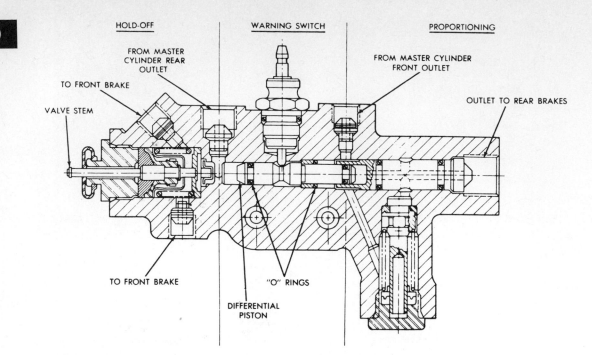

Figure 3-65. A combination brake valve for a front-rear split hydraulic system includes a hold-off (metering) valve, a brake warning light switch (pressure differential switch), and a proportioning valve. Many cars equipped with disc and drum brakes use this valve.

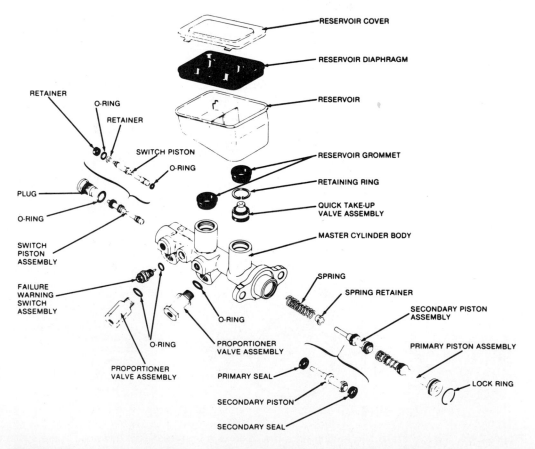

Figure 3-66. Quick take-up master cylinder assembly for a diagonally split hydraulic brake system used on a front-wheel-drive vehicle. Note two separate proportioning valves (one for each rear wheel), differential switch piston, and failure switch. *(Courtesy of General Motors Corporation)*

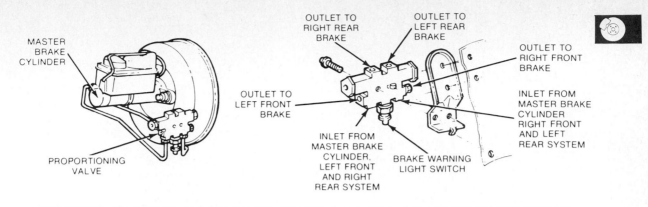

MASTER BRAKE CYLINDER

PROPORTIONING VALVE

OUTLET TO RIGHT REAR BRAKE

OUTLET TO LEFT REAR BRAKE

OUTLET TO RIGHT FRONT BRAKE

INLET FROM MASTER BRAKE CYLINDER RIGHT FRONT AND LEFT REAR SYSTEM

OUTLET TO LEFT FRONT BRAKE

INLET FROM MASTER BRAKE CYLINDER, LEFT FRONT AND RIGHT REAR SYSTEM

BRAKE WARNING LIGHT SWITCH

Figure 3-67. Combination valve-and-switch assembly for one type of diagonally split brake hydraulic system. This unit includes the pressure differential valve and switch and proportioning valves. *(Courtesy of Ford Motor Co. of Canada Ltd.)*

Pressure Differential Switch

(Warning Light)

A brake warning light switch is used with the dual braking system. When a hydraulic leak develops in either system and the brake pedal is applied, a pressure difference is sensed by the switch completing the electrical circuit to the light and causing it to go on. This warns the driver that the brake system needs attention. An additional function for this brake light on many cars is to warn the driver that the parking brake is applied and should be released before starting to drive.

Metering Valve

Cars equipped with disc brakes for the front-rear split system sometimes have a metering valve in the front-brake hydraulic system. This prevents the front brakes from applying until the rear brakes are applying at approximately 90 to 180 psi (620.55 kilopascals to 1241.1 kilopascals). Since front disc brakes have no brake shoe return springs for the hydraulic pressure to overcome, they would apply too soon and could cause loss of directional control of the car. The diagonally split system requires two metering valves.

Proportioning Valve

During severe braking, more of the car weight is transferred to the front wheels. As a result, the front wheels do most of the braking. Without a proportioning valve, the rear wheels, having relatively little traction, would skid. The proportioning valve reduces hydraulic pressure buildup to the rear wheels as compared to pressure buildup at the front wheels after a specified pressure has been reached (approximately 350 psi or 2,413.25 kilopascals). In the diagonal split system, two proportioners are required---one for each system.

Combination Valve

The combination valve combines the brake warning light switch, metering valve, and proportioning valve into one unit. Many cars with the front-rear split system use this type of valve.

Brake Fluid

The fluid used in the hydraulic brake system is a special fluid, and only brake fluid of high quality should be used. Mineral oil, hydraulic oil, automatic transmission fluid, power steering fluid, or other oils should *never* be used in a hydraulic brake system.

High-quality brake fluid has a number of characteristics. It is able to absorb moisture and therefore should never be left open to atmosphere, since the atmosphere contains moisture. It has a high boiling point to prevent evaporation, which would result in loss of braking effectiveness. A boiling point of 550°F. (287.8°C.) or higher is needed for present brake systems. Good brake fluid is noncorrosive, has good lubricating qualities, remains stable over a long period of time, and is compatible with other high-quality brake fluids. Brake fluid must be kept absolutely clean and in airtight containers to maintain its effectiveness.

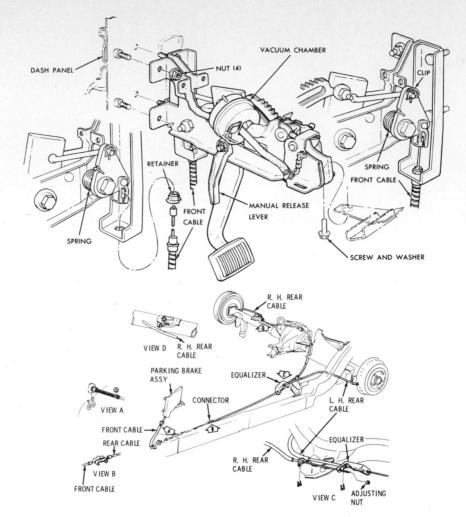

Figure 3-68. Parking brake linkage. The equalizer provides equal apply force to both rear wheels. Also shown is a vacuum-operated parking brake automatic release. *(Courtesy of General Motors Corporation)*

PART 5   PARKING BRAKES

A parking brake may be operated by a separate foot pedal or by a hand-operated lever. The lever or pedal operates a cable system linked to the rear brakes and applies them mechanically. Pushing down the pedal tightens the cable and forces the shoes against the drum. Some cars use a vacuum motor to release the parking brake automatically as soon as the car starts. In four-wheel disc brakes two different types of parking brake are used as described earlier. Older cars and some trucks use a drum mounted at the rear of the transmission for the parking brake.

PART 6   POWER BRAKES

More and more cars are equipped with power brakes. Power brakes reduce the amount of force the driver needs to apply to the brake pedal. Two types of power brakes are in use. The vacuum-operated power brake is the most common. The other type of power brake is hydraulically operated from power steering pump pressure. This unit is commonly called Hydro-Boost or Hydro Max.

Vacuum Power Brake

The vacuum-suspended power brake uses the engine intake manifold vacuum. A vacuum booster is mounted between the master cylinder and the firewall. The brake pedal pushes against the booster pushrod, which in turn operates a valve assembly. The valve assembly closes off the vacuum to the rear chamber of the booster and admits atmospheric pressure to that side. This causes the piston to move toward the master cylinder. The piston pushes against the master cylinder pushrod, thereby providing braking assistance. Most of the apply force is provided by the booster; however, a *reaction* assembly allows part of the apply pressure to be felt by the driver, giving the driver needed pedal *feel*.

The booster can be in the *released, apply*, or *hold* position. In the released position the atmospheric valve is closed and the intake manifold vacuum is present on both sides of the piston, allowing the piston return spring to hold the piston in the release position. When the brake pedal is depressed, the booster is in the apply position. The vacuum port to the rear booster chamber is closed, and the atmospheric port is opened. This allows atmospheric pressure to push the piston forward, applying the brakes. As soon as pedal movement stops, the booster is in the hold position. In this position the piston has caught up with pedal and pushrod movement, causing the atmospheric port to close. Since the vacuum port is already closed, the booster holds the apply position desired by the driver. As soon as the pedal is released, the atmospheric port closes and the vacuum port opens, allowing the spring to return the booster piston to the release position.

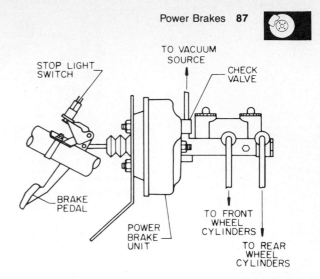

Figure 3-69. Power brake and master cylinder mounting arrangement. *(Courtesy of General Motors Corporation)*

The vacuum power brake stores sufficient vacuum for several brake applications should the engine fail to provide vacuum. Heavier cars use a dual-piston booster for added braking assist.

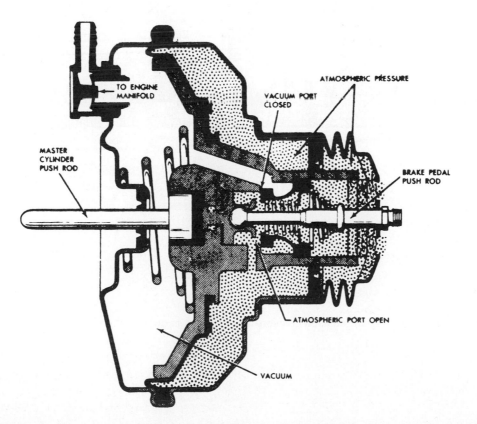

Figure 3-70. Vacuum brake booster in apply position. Vacuum port is closed and atmospheric port is open.

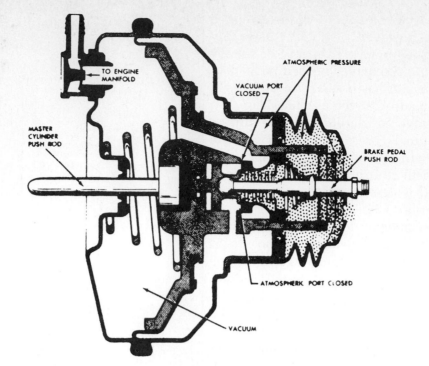

Figure 3-71. Booster in hold position. Both ports are closed.

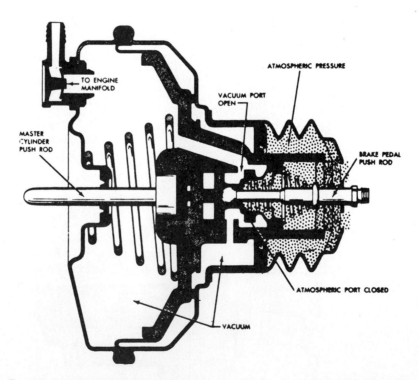

Figure 3-72. Booster in released position. Vacuum port is open and atmospheric port is closed.

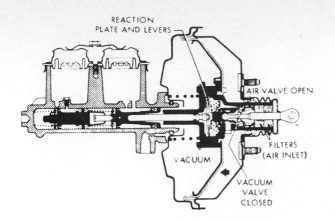

Figure 3-73. Delco-Moraine lever-type reaction mechanism.

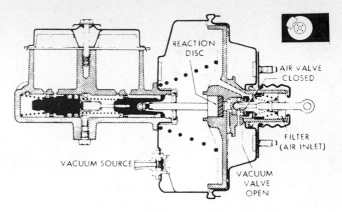

Figure 3-74. This Bendix booster uses a disc-type reaction.

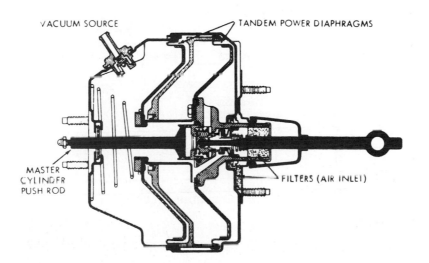

Figure 3-75. This tandem power diaphragm booster is used on larger cars requiring more apply pressure, since a heavier car requires more braking effort to bring it to a stop.

Hydro-Boost Power Brake

The addition of emission-control devices to the engine reduced the available intake manifold vacuum, and in some cases there was insufficient vacuum to operate a vacuum power brake. A hydraulic brake booster called Hydro-Boost was used to overcome this problem. This unit uses power-steering pump hydraulic pressure to assist in applying the brakes.

When the brake pedal is depressed, a control valve in the Hydro-Boost unit allows power-steering pump pressure to act on a piston, which in turn pushes on the master cylinder pushrod. The amount of assist depends on pedal pressure and the position of the control valve. In the hold position the valve shuts off any additional hydraulic pressure from the power-steering pump to the booster. On release, the valve dumps apply pressure as the piston is returned to the release position.

An accumulator is added to store sufficient hydraulic pressure for several brake applications should the power-steering pump fail. After that, the brakes operate normally but without power assist.

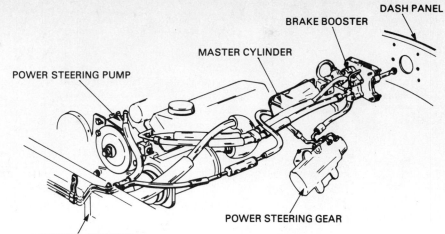

Figure 3-76. Hydro-Boost power brake system layout. (*Courtesy of Ford Motor Co. of Canada Ltd.*)

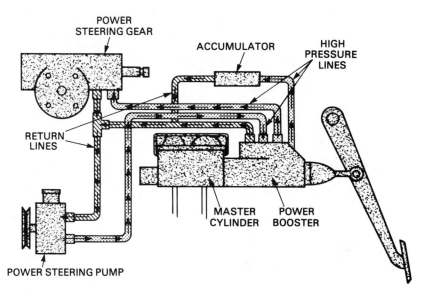

Figure 3-77. Hydro-Boost fluid flow diagram.

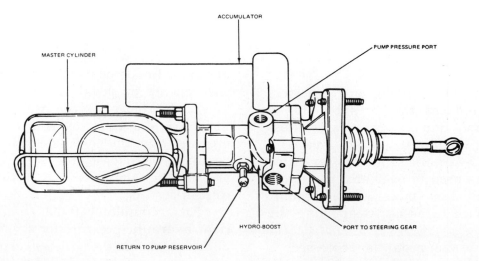

Figure 3-78. Hydro-Boost with master cylinder attached. (*Courtesy of Ford Motor Co. of Canada Ltd.*)

Diagnostic Chart

PROBLEM	CAUSE	CORRECTION
Pedal bottoms out; no brakes	1. Fluid leak. 2. Air in system. 3. Lining worn. 4. Low fluid level. 5. Linkage disconnected. 6. Automatic shoe adjusters not functioning. 7. Vaporized fluid.	1. Repair source of leak. 2. Bleed system. Repair source of air entry. 3. Adjust or reline. 4. Fill and bleed system. 5. Connect. 6. Replace adjusters. Adjust shoes. 7. Install super-heavy-duty fluid and bleed.
Spongy pedal	1. Air in system. 2. Drums too thin. 3. Soft hose. 4. Shoe lining wrong fit. 5. Cracked brake drum. 6. Brake shoes distorted.	1. Bleed system. Repair source of air entry. 2. Replace drums. 3. Replace hose. 4. Install correct lining. 5. Replace drum. 6. Replace shoes.
Hard pedal; little braking	1. Incorrect lining. 2. Linings contaminated. 3. Primary and secondary shoes reversed. 4. Brake linkage binding. 5. Master or wheel cylinder pistons frozen. 6. Linings hard and glazed. 7. Lining ground to wrong radius. 8. Brake line or hose clogged or kinked. 9. Power booster unit defective. 10. No vacuum to power booster. 11. Engine fails to maintain proper vacuum to booster.	1. Install proper lining. 2. Replace or reline shoes. Repair source of leak. 3. Install shoes in correct location. 4. Free and lubricate. 5. Rebuild or replace cylinder. 6. Replace linings. 7. Grind lining as specified. 8. Replace. 9. Replace power booster. 10. Replace clogged, soft lines. Repair leaks. 11. Tune or overhaul engine.
Pedal fade	1. Excessive use of brakes. 2. Poor brake fluid. 3. Improper lining to drum contact. 4. Thin brake drums. 5. Dragging brakes. 6. Riding the brake pedal.	1. Use lower gears, reduce speed, load, etc. 2. Flush. Install super-heavy-duty fluid. 3. Adjust shoes or grind to correct radius. 4. Install new drums. 5. Adjust or repair other cause of dragging. 6. Keep foot from brake unless needed.
Pulsating pedal	1. Brake drums out of round. 2. Excessive disc run-out. 3. Loose wheel bearings. 4. Drums loose. 5. Rear axle bent.	1. Turn drums in pairs. 2. Replace disc or recondition. 3. Adjust. 4. Tighten wheel lugs. 5. Replace axle.
Brakes grab	1. Grease or brake fluid on lining. 2. Lining charred. 3. Lining loose on shoe. 4. Loose wheel bearings. 5. Defective wheel bearings. 6. Loose brake backing plate. 7. Defective drum. 8. Sand or dirt in brake shoe assembly. 9. Wrong lining. 10. Primary and secondary shoes reversed.	1. Install shoes, correct leak. 2. Reline. 3. Replace. 4. Adjust. 5. Replace. 6. Torque fasteners. 7. Turn drum. Turn drum on opposite side also to same size. 8. Disassemble and clean. 9. Install correct lining. 10. Install correctly.
Car pulls to one side, brakes applied	1. One wheel grabbing. 2. Different lining on one side or shoes reversed on one side. 3. Plugged line or hose. 4. Uneven tire pressure. 5. Front end alignment out. 6. Sagged, weak, or broken spring. Weak shock absorber.	1. See "Brakes grab." 2. Replace lining or install shoes in proper position. 3. Replace. 4. Correct pressures. 5. Align front end. 6. Install new spring or shocks.
Brakes drag	1. Parking brake too tight. 2. Clogged hose or line.	1. Adjust. 2. Replace.

PROBLEM	CAUSE	CORRECTION
	3. No pedal free travel.	3. Adjust pedal free travel so that compensating port will be open when brake is released.
	4. Brakes adjusted too tight.	4. Adjust.
	5. Master cylinder or wheel cylinder cups soft and sticky.	5. Rebuild or replace cylinders. Flush system.
	6. Loose wheel bearing.	6. Adjust.
	7. Parking brake fails to release.	7. Clean and lubricate parking brake linkage and adjust.
	8. Shoe retracting springs weak or broken.	8. Replace.
	9. Out-of-round drum.	9. Turn drum in pairs.
	10. Defective power booster.	10. Replace booster.
Brakes chatter	1. Weak or broken shoe retracting springs.	1. Replace.
	2. Defective power booster.	2. Replace booster.
	3. Loose backing plate.	3. Tighten.
	4. Loose or damaged wheel bearings.	4. Adjust or replace bearings.
	5. Drums tapered or barrel shaped.	5. Turn drum in pairs.
	6. Bent shoes.	6. Replace shoes.
	7. Dust on lining.	7. Clean.
	8. Lining glazed.	8. Replace.
	9. Drum dampener spring missing.	9. Install dampener spring.
	10. Grease or fluid on linings.	10. Reline brakes and correct leak.
	11. Shoes not adjusted properly.	11. Adjust.
Brakes squeal	1. Glazed or charred lining.	1. Replace.
	2. Lining rivets loose.	2. Replace.
	3. Wrong lining.	3. Replace with correct lining.
	4. Shoe hold-downs weak or broken.	4. Replace.
	5. Drum damper spring missing.	5. Install spring.
	6. Shoes improperly adjusted.	6. Adjust to specifications.
	7. Shoes bent.	7. Replace.
	8. Bent backing plate.	8. Replace plate.
	9. Shoe retracting springs weak or broken.	9. Replace springs.
	10. Drum too thin.	10. Replace drum.
	11. Lining saturated with grease or brake fluid.	11. Replace linings.
Brake shoes click	1. Shoe is pulled from backing plate by following tool marks in drum.	1. Machine drum properly, in pairs.
	2. Shoe bent.	2. Replace.
	3. Shoe support pads on backing plate grooved.	3. Smooth and lubricate pads or replace.

General Precautions

1. The importance of high-quality workmanship on brake repairs cannot be overemphasized. The lives of the driver, the passengers, and other people on the road depend on the ability of the car's brake system to bring it to a safe, controlled stop. Shoddy or inferior workmanship can cause injury and loss of life.

2. Safe procedures must also be followed when raising a car on a hoist or on stands to do brake work. Never work on a car supported only on a jack. Place the car on a hoist or place stands where recommended by the vehicle manufacturer.

3. Absolute cleanliness must be observed when servicing hydraulic brake parts. Hydraulic parts must be properly cleaned in recommended brake cleaning fluid and kept clean. During assembly, these parts should be lubricated with clean, new brake fluid or other recommended assembly lubricant. Friction surfaces such as brake linings, pads, drums, and discs must not have any grease, oil, or brake fluid in contact with them. Contamination of this kind destroys their frictional characteristic and can cause grab and pull.

4. When lubricating brake mechanisms inside the brake drum unit, such as backing plates and adjusters, do not overlubricate. Overlubrication can cause contamination of linings and drums and result in grab or pull. Use only recommended high-temperature lubricant.

5. Brake fluid should not be allowed to come in contact with painted surfaces since it is a very effective paint remover. If fluid is accidentally spilled onto a painted surface, it should immediately be washed thoroughly with water.

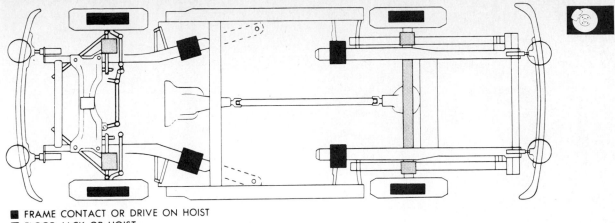

■ FRAME CONTACT OR DRIVE ON HOIST
▨ FLOOR JACK OR HOIST
○ BUMPER JACK (AT BUMPER SLOT ONLY)

Figure 3-79. Hoisting or jacking points are shown for one particular type of vehicle. Other vehicles may have different lift points.

Disassembly Procedure (Typical Dual-Servo Disc-Drum Brakes)

1. Place the car on stands. Position stands safely where recommended by the manufacturer. If work is to be done with the car on a hoist, position the car on the hoist and lift at the points recommended by the vehicle manufacturer.

2. Remove wheel mounting studs or nuts. (Some older cars have left-hand thread on the left side.) Mark the wheels for proper assembly.

3. Remove the brake drums. Some drums have speed nut retainers or a small screw holding drum to axle flange or hub. If the drum is worn, the brake shoe adjustment will have to be backed off for the drum to clear the brake shoes. Do not force the drum or distort it. Do not allow the drum to drop.

4. Remove caliper mounting bolts, brake line, calipers, and pads.

5. Remove front-wheel dust cap, cotter pin, and a wheel bearing nut from the front wheels. A slight wiggle of the brake disc by hand will cause the outer wheel bearing to slide out. Remove the disc from the spindle.

6. Remove brake shoe return springs, hold-down springs and pins, adjuster, and brake shoes from rear-wheel backing plates. Do not distort or overstretch the springs.

7. Remove the wheel cylinders by disconnecting brake line and wheel cylinder mounting bolts.

8. Remove the master cylinder (first disconnect the brake pedal linkage on nonpower brakes). Then disconnect brake lines, master cylinder mounting nuts, and master cylinder.

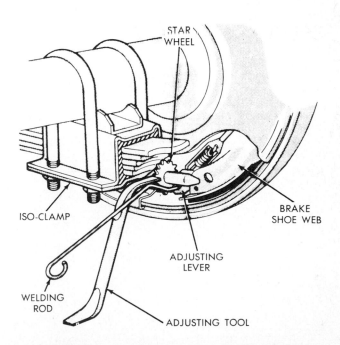

Figure 3-80. Manual adjustment on self-adjusting brakes. This must be backed off for drum removal when drums are worn.

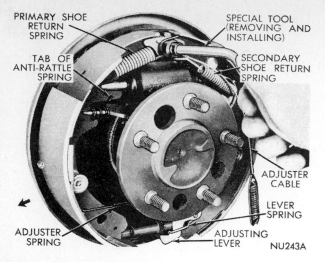

PRIMARY SHOE
RETURN
SPRING

TAB OF
ANTI-RATTLE
SPRING

SPECIAL TOOL
(REMOVING AND
INSTALLING)

SECONDARY
SHOE RETURN
SPRING

ADJUSTER
CABLE

LEVER
SPRING

ADJUSTER
SPRING

ADJUSTING
LEVER

NU243A

Figure 3-81. Removing brake shoe retracting springs with special tool.

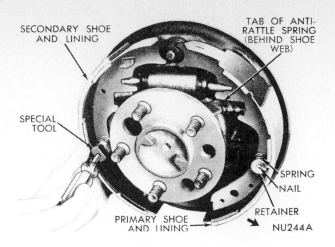

SECONDARY SHOE
AND LINING

TAB OF ANTI-
RATTLE SPRING
(BEHIND SHOE
WEB)

SPECIAL
TOOL

SPRING

NAIL

RETAINER

PRIMARY SHOE
AND LINING

NU244A

Figure 3-82. Removing brake shoe hold-down springs.

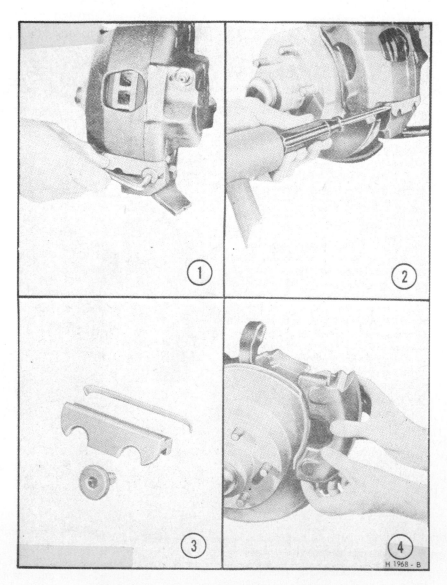

H 1968 - B

Figure 3-83. Removal of Ford type of sliding caliper.

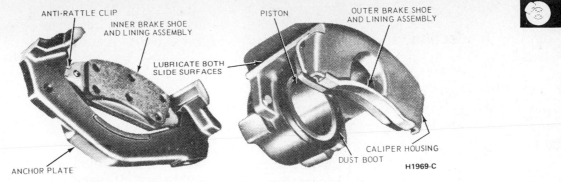

Figure 3-84. Caliper and anchor plate after removal.

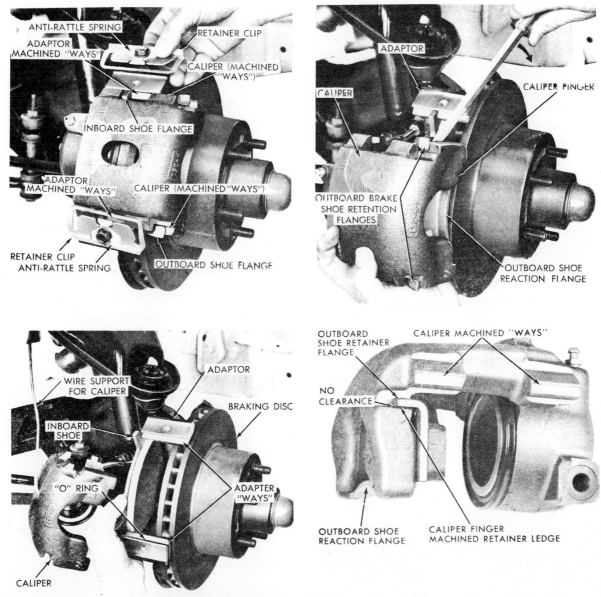

Figure 3-85. Removal of Chrysler's sliding caliper.

Cleaning, Inspection, and Measuring Procedure

1. Check *rear axle seals* for leakage. If leaking differential oil, axle shafts must be removed, seals and bearings replaced, and axle shafts installed.

2. Clean *backing plates* with parts-cleaning solvent. Blow dry with compressed air. Inspect ledges for wear. If not much wear is present, sand ledges smooth. Inspect anchor pin for wear. Inspect backing plate for damage. Plate should not be bent or physically damaged.

3. Inspect *brake shoe and linings* Uneven lining wear can indicate improperly mounted parts, drum distortion, or wear. Linings should be replaced before lining wear allows metal rivets or brake shoe metal to contact the drum. Some disc brake pads have a wear indicator telling the driver or mechanic when they should be replaced. New linings must be fitted to the drum with 0.006- to 0.010-inch (0.1524 to 0.254 millimeter) heel-and-toe clearance on most servo-type brakes. If linings do not have this kind of heel-and-toe clearance, they must be re-arced to fit.

4. Clean and inspect all *springs, linkages,* and *adjusters* for wear or distortion. Springs should not be discolored from overheating and should not be stretched or distorted. Adjuster threads and sockets should be cleaned and lubricated with a good-grade brake lubricant. Replace any faulty parts.

5. Measure *drums* for wear and out-of-round. If drum wear exceeds the manufacturer's limit, drums must be replaced. In most cases, for car brake drums the maximum allowable oversize after machining is 0.060 inch (1.524 millimeter) over standard drum diameter. The manufacturer's specifications must be followed for maximum safe tolerances. If wear is within limits, drums should be machined to recondition the friction surface. If machining the drums removes enough material to increase drum diameter beyond the manufacturer's specifications, the drum must be replaced. Brake drums should not have excessive run-out (wobble) or out-of-round. Follow the manufacturer's specifications.

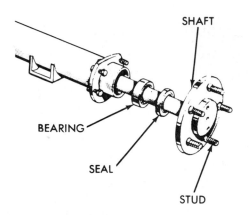

Figure 3-87. Location of rear axle seal, which must be replaced if differential gear oil leaks into brake assembly.

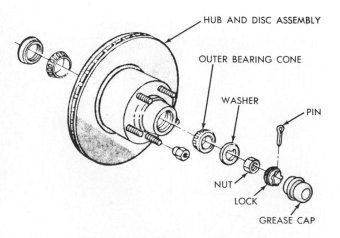

Figure 3-86. Disc and wheel bearings after removal.

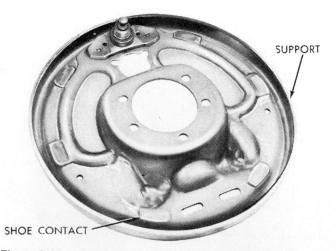

Figure 3-88. Backing plate after removal, showing shoe contact platforms or ledges.

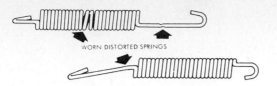

Figure 3-89. Worn or distorted springs should be replaced.

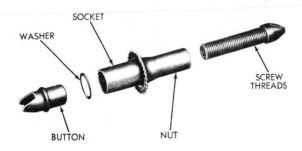

Figure 3-90. Star-wheel adjuster exploded view.

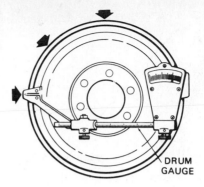

Figure 3-91. Brake drum must be measured at several points for wear and out-of-round.

Figure 3-92. Most recent brake drums show the maximum allowable drum diameter stamped on the drum.

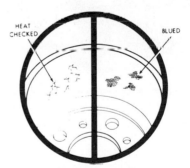

Figure 3-93. Drums that are overheated show heat checks and blued spots that have become hardened. This may prevent machining.

6. Inspect *discs* for scoring and wear. Severe scoring requires disc replacement. Minor scoring can be corrected by machining. Measure the discs for thickness. Discs should not be less than manufacturer's stated thickness limit usually stamped on disc. Minimum thickness refers to rotor or disc thickness after machining. The discs should also be measured for run-out, surface parallelism, and thickness variation. Usual tolerance is 0.0005 inch (0.0127 millimeter) for parallelism and thickness variation and .002 inch (.051 millimeter) to .005 inch (.127 millimeter) for run-out. Follow the manufacturer's recommendation. Front-wheel hubs should be cleaned thoroughly after the bearings are removed.

7. Clean and inspect *front-wheel bearings* as described in Chapter 2. Wash bearings in solvent and blow dry with compressed air. *Caution:* Do not spin bearings with compressed air. This can damage bearings, and they may fly apart, causing injury. Inspect bearing rollers for damage or discoloration. Bearing roller cage should not be distorted or bent. Bearing cups should show a normal, dull-gray wear pattern. Replace damaged, worn, or rough bearings.

8. Disassemble the wheel cylinder. Inspect the cylinder bore for minor damage. If the cylinder bore is not excessively corroded, rusted, or worn, it can be honed and a new kit installed. Check piston-to-bore clearance after honing. Clearance should not exceed the manufacturer's specifications, usually 0.003 to 0.004 inch (0.0762 to 0.1016 millimeter). Wash the cylinder thoroughly in brake cleaning fluid and assemble with parts from new kit. Parts should be lubricated with lubricant provided with the kit or with clean, new brake fluid. Make sure that the bleeder screw is free, clear, and in good condition.

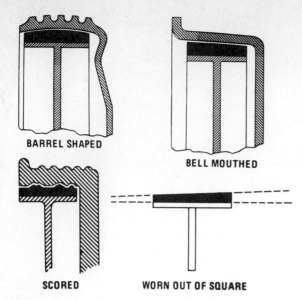

BARREL SHAPED

BELL MOUTHED

SCORED

WORN OUT OF SQUARE

Figure 3-94. Other drum conditions that may require machining or drum replacement.

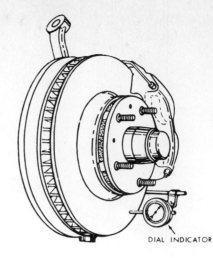

DIAL INDICATOR

Figure 3-97. Using a dial indicator to measure rotor runout.

MINIMUM THICKNESS MARKING

Figure 3-95. Rotor showing minimum thickness marking stamped on the disc.

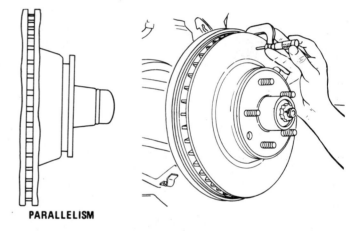

PARALLELISM

Figure 3-98. Measuring rotor thickness and parallelism with a micrometer.

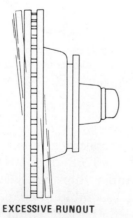

EXCESSIVE RUNOUT

Figure 3-96. Rotor run-out causes piston "knock back" with resultant loss of pedal reserve, as well as a pulsating brake pedal.

9. Disassemble the *master cylinder* by removing the primary piston retainer snap ring or screw and the secondary piston stop screw (if so equipped). The secondary piston stop screw is located inside the reservoir on some master cylinders. Remove the pistons and springs. Examine the cylinder bore condition. Minor damage can be polished out with crocus cloth or with a brake cylinder hone. If corrosion or rust is excessive, the master cylinder must be replaced. During honing no more than just a few thousandths of an inch should be removed from the cylinder to maintain proper piston-to-bore clearance. Measure with a feeler gauge between piston and bore. Follow the manufacturer's specifications for

maximum allowable clearance. Honing is not recommended for some types of master cylinders. After honing and after checking piston-to-bore clearance, throroughly wash the master cylinder in brake cleaning fluid and assemble with the new master cylinder kit. A rebuild kit usually contains the primary and the secondary parts, as well as residual check valves and seats where applicable. Lubricate all internal parts with clean, new brake fluid. The master cylinder is then ready for installation.

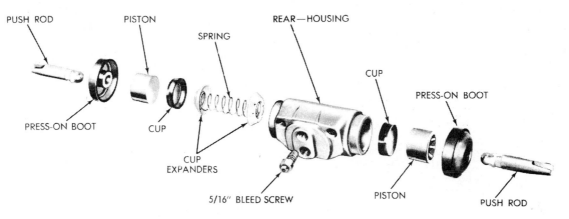

Figure 3-99. Exploded view of common type of wheel cylinder.

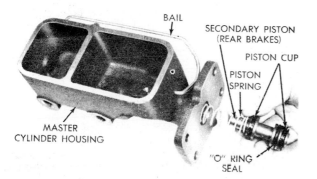

Figure 3-100. Removal or installation of secondary piston in dual master cylinder.

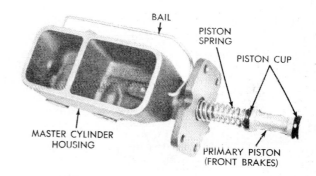

Figure 3-101. Removal or installation of master cylinder primary piston in dual master cylinder.

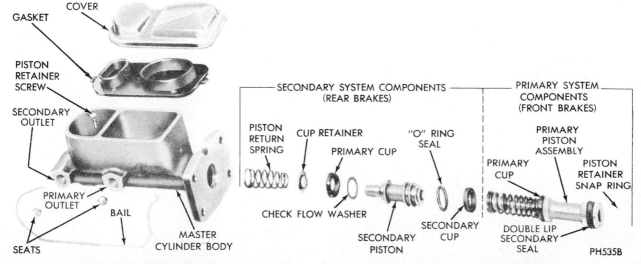

Figure 3-102. Exploded view of tandem master cylinder component parts disassembled.

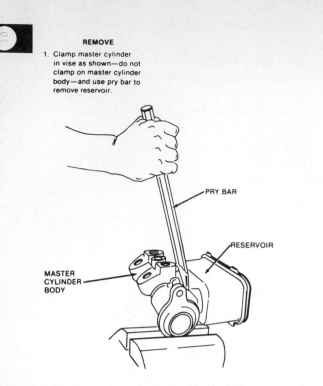

REMOVE

1. Clamp master cylinder in vise as shown—do not clamp on master cylinder body—and use pry bar to remove reservoir.

PRY BAR

RESERVOIR

MASTER CYLINDER BODY

INSTALL

1. Lay reservoir on flat, hard surface as shown. Press on master cylinder body using rocking motion.

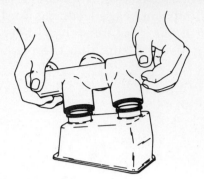

Figure 3-104. Installing master cylinder reservoir. *(Courtesy of General Motors Corporation)*

Figure 3-103. Removing plastic master cylinder reservoir. *(Courtesy of General Motors Corporation)*

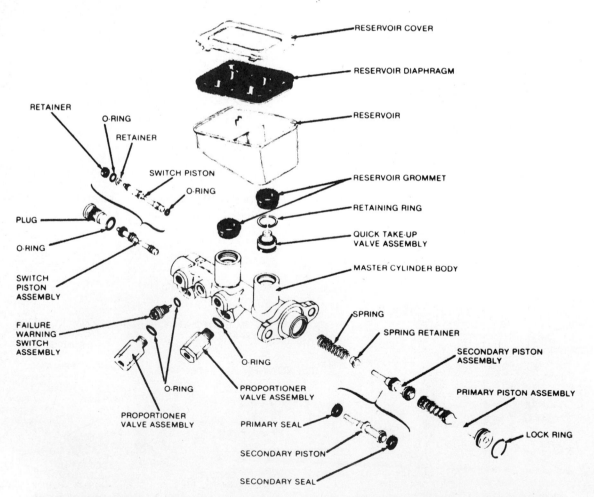

RESERVOIR COVER

RESERVOIR DIAPHRAGM

RESERVOIR

RESERVOIR GROMMET

RETAINING RING

QUICK TAKE-UP VALVE ASSEMBLY

MASTER CYLINDER BODY

SPRING

SPRING RETAINER

SECONDARY PISTON ASSEMBLY

PRIMARY PISTON ASSEMBLY

LOCK RING

RETAINER

O-RING

RETAINER

SWITCH PISTON

O-RING

PLUG

O-RING

SWITCH PISTON ASSEMBLY

FAILURE WARNING SWITCH ASSEMBLY

O-RING

PROPORTIONER VALVE ASSEMBLY

O-RING

PROPORTIONER VALVE ASSEMBLY

PRIMARY SEAL

SECONDARY PISTON

SECONDARY SEAL

Figure 3-105. Typical master cylinder component parts for diagonally split brake hydraulic system. This master cylinder has a plastic reservoir and an aluminum die cast housing. *(Courtesy of General Motors Corporation)*

10. *Lines,* *switches,* and *valves* should be inspected for damage. Flex lines should not be cracked or swollen. Steel lines should be checked for dents and rust. Severely damaged lines should be replaced with seamless steel brake lines of the same diameter. Lines must be bent to fit with the proper bending tool to avoid kinking or restricting the line. The metering valve can be tested by installing a pressure gauge between the valve and the front disc brakes. By operating the brake pedal, the opening pressure of the metering valve can be observed. An alternate method of testing the metering valve is to depress the brake pedal carefully. A slight "bump" should be felt as the brake pedal passes the first inch of pedal travel. This indicates that the metering valve is working.

To test the proportioning valve, a pressure gauge must be installed on each side of the valve. Operating the brake pedal will show pressure on both sides of the valve when the *split* point has been reached. When the gauge between the master cylinder reads 500 psi (3,447.5 kPa), the other gauge should read approximately 350 psi (2,413.25 kPa). If the brake warning light stays on, it may be because the switch is not the self-centering type. This often happens after bleeding brakes. To correct this condition, bleed the side opposite to the one that was bled last. For example, if the primary side was bled last, bleed the secon-

dary side until the light just goes off. This will center the switch and turn the light off. To check if the bulb is working, watch the warning light when the switch is turned to the start position; it should light. If not, replace the bulb or check the circuit.

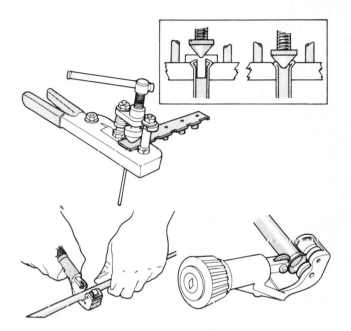

Figure 3-107. Sequence for flaring steel lines with a double-lap flare. *(Courtesy of Chrysler Corporation)*

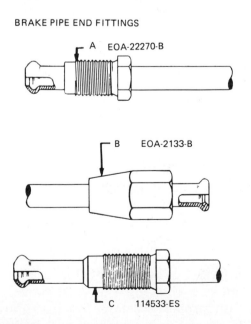

Figure 3-106. Several types of brake line fittings. Note double-lap flared tubing at center and ISO-type tubing at top and bottom.

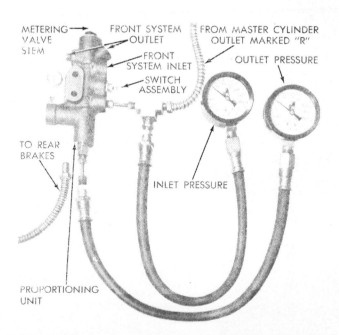

Figure 3-108. Gauge connections for testing proportioning valve operation. At 500 psi, master cylinder pressure proportioning valve output pressure should be approximately 350 psi. If defective, valve must be replaced.

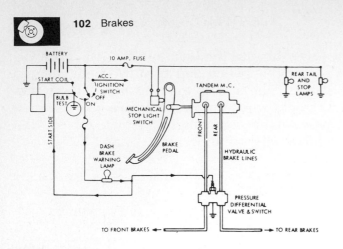

Figure 3-109. Hydraulic system schematic showing pressure differential switch and brake warning light.

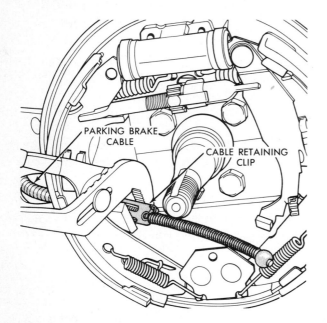

Figure 3-110. To remove parking brake cable from backing plate, squeeze retaining clip with pliers and slide cable through plate. *(Courtesy of Chrysler Corporation)*

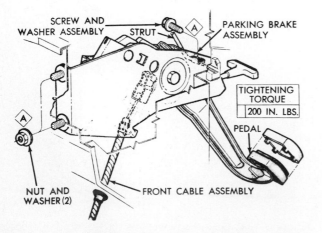

Figure 3-111. Parking brake pedal mechanism showing cable connection. *(Courtesy of Chrysler Corporation)*

11. Inspect the *parking brake* linkage and cables for damage. Damaged parts should be replaced. Cables and linkage should operate freely and be properly lubricated to assure continued good operation. Adjust the parking brake according to the manufacturer's specifications. This is usually done at the equalizer between the front and rear cables, after the service brake adjustment.

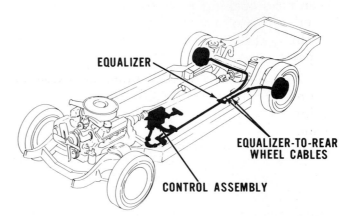

Figure 3-112. Location of parking brake equalizer. Adjustment is usually made at this point. *(Courtesy of Ford Motor Co. of Canada Ltd.)*

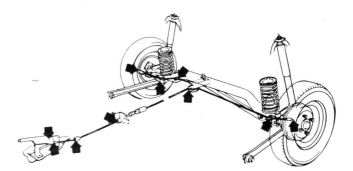

Figure 3-113. Lubrication points for typical parking brake control mechanism. *(Courtesy of Ford Motor Co. of Canada Ltd.)*

Machining and Reconditioning Procedure

1. Brake drums and discs must be machined to restore balanced and effective braking. Always machine drums and discs in axle pairs. When machining drums, start with the one showing the most wear. If it cleans up properly before the maximum allowable diameter is reached, it can be used again. Machine the other drum to within

0.010 inch (0.254 millimeter) of this diameter to prevent pull to one side. Follow the machine manufacturer's instructions for mounting and machining procedures. Brake discs need not be machined to the same thickness, since a thickness difference in two front discs will not cause uneven braking or pull. When machining discs, remove as little as possible and still achieve the desired friction surfaces. Discs should be replaced if they do not recondition by the time minimum thickness has been reached. Follow the equipment manufacturer's instructions for mounting discs on the machine and for machining. Always machine in pairs.

2. Brake shoe linings must be fitted to the new or reconditioned drum to provide full lining-to-drum contact during braking and to avoid squeal and chatter. This requires arcing or radius grinding of the lining to provide

from 0.006- to 0.010-inch (0.1524 to 0.253 millimeter) heel-and-toe clearance when the shoe is held in the drum by hand. It is good practice to chalk the the entire lining surface before radius grinding. This makes it easier to see when the entire surface has been ground. Remove as little material as possible when radius grinding and still achieve the desired fit. Follow the equipment manufacturer's instructions for mounting shoes in machine and for proper adjustment and grinding procedure.

3. *Wheel cylinder, master cylinder,* and *caliper piston bores* are reconditioned in a similar manner. Minor nicks, scratches, and burrs should be dressed with crocus cloth. Anything more serious requires honing of the cylinder, where the manufacturer allows this procedure. If the bore diameter is increased more than the permissible amount before the cylinder cleans up, the master cylinder, wheel cylinder, or caliper must be replaced. Honing is usually followed by polishing with crocus cloth to ensure a smooth cylinder surface and to prevent damage to cups and seals. Thorough cleaning of these parts is necessary after honing to insure that no abrasive particles are allowed to enter the hydraulic system. Pistons can also be polished with crocus cloth, after which they should be thoroughly cleaned. All new internal rubber parts, pistons, and bores should be thoroughly lubricated with clean, new brake fluid or recommended special as-

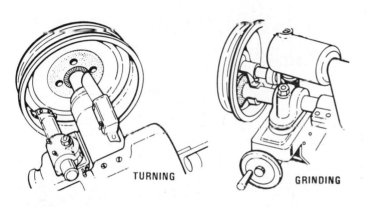

Figure 3-114. Brake drum mounted for machining on two different types of machines.

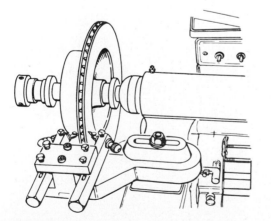

Figure 3-115. Disc brake rotor mounted in machine for resurfacing.

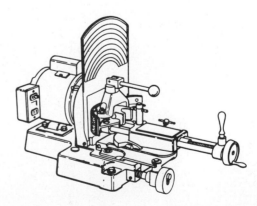

Figure 3-116. Brake shoe contour grinding is required to provide proper lining to drum-fit.

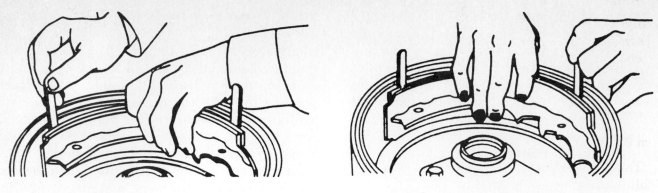

.010' CLEARANCE AT BOTH ENDS OF LINING

Figure 3-117. Checking heel-and-toe clearance with feeler gauges.

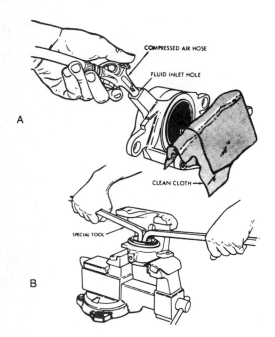

Figure 3-118. Removing caliper piston with air pressure (A) and with special tool (B).

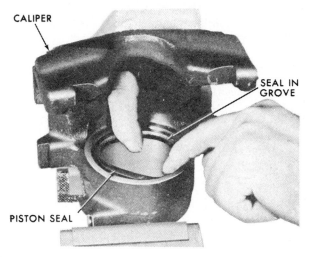

Figure 3-120. Installing the piston seal in caliper piston bore.

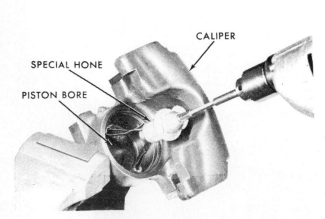

Figure 3-119. Honing a caliper piston bore.

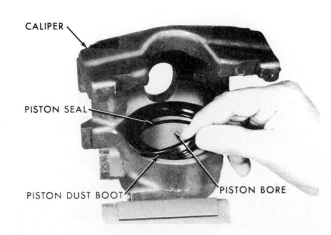

Figure 3-121. Installing the caliper piston boot after seal and piston have been installed.

sembly lubricant during assembly. Avoid using screwdrivers on rubber seals and boots to prevent puncturing or cutting. A wooden or plastic tool will avoid this.

Assembly and Adjustment Procedure

Drum Brake Unit

To assemble the drum brake unit, proceed as follows.

• Lubricate the parking brake at all points recommended in the service manual.

• Mount the backing plate on the spindle at the front and the axle flange at the rear. Install the rear axle seal and axle shaft.

• Lubricate the shoe platforms on the backing plate. Use a good-quality, high-temperature lubricant that will not run, and use it sparingly to prevent lining contamination.

• Mount the assembled wheel cylinder and attach the brake line.

• Attach the parking brake cable to the secondary shoe lever at the rear.

• Mount the shoes, hold-downs, parking brake link, and adjuster mechanism.

• Make sure that the parking brake is fully released so that it will not interfere with service brake adjustment.

• Make sure that the primary and secondary shoe retracting springs are not interchanged.

• The completed shoe assembly should float or slide freely on the backing plate.

• Use the gauge to determine proper shoe adjustment and adjust shoes to drum size.

• Install the brake drum and drum retainer if used.

• Install wheel and torque bolts or nuts in proper sequence to specifications.

• Repack wheel bearings with wheel bearing grease as described in Chapter 2.

• *Caution*: Be sure not to touch the drum or disc friction surface with grease, greasy hands, or brake fluid. Friction surfaces must remain clean and dry.

Disc Brake Unit

After wheel bearings have been repacked and the disc installed, proceed as follows.

• Install new disc brake pads on a new or reconditioned caliper.

• Make sure that any anti-rattle clips and materials are in place with pads.

• The caliper piston must be forced back into the caliper to provide sufficient room between the pads to fit over the disc. Use a spreader between the pads or a C-clamp to do this.

• Mount the caliper over the disc.

• Make sure that all attaching parts, bushings, clips, and retainers are properly positioned. Tighten mounting bolts to specified torque. Connect the brake line.

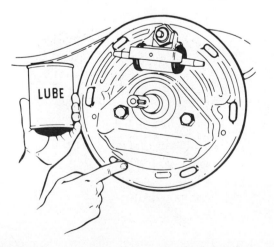

Figure 3-122. Lubricating backing plate ledges with special lubricant. Avoid overlubrication.

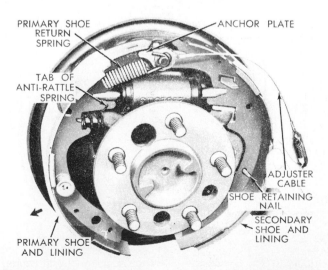

Figure 3-123. Drum brake assembly nearly completed. Primary (short) lining must be at front of assembly. Primary and secondary return springs must not be interchanged.

• Attach the wheel and torque to specifications in proper sequence.

Master Cylinder

Attach the master cylinder to the firewall or power brake unit. Be sure that the master cylinder piston returns to the fully released position. This requires proper pushrod ad-

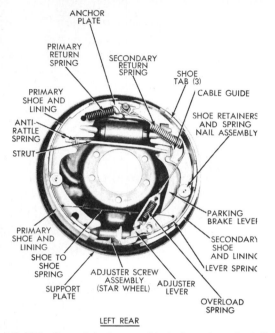

Figure 3-124. Completed assembly before drum installation should be checked for free movement on backing plate. Self-adjuster operation can be checked by tightening cable by hand. When cable is released, the adjuster should turn, slightly expanding the brake shoes.

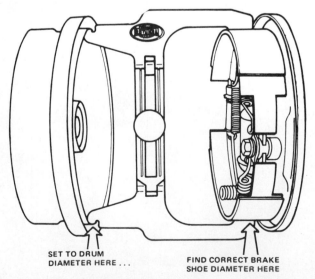

Figure 3-125. Checking brake shoe adjustment with gauge. By using this method, proper lining-to-drum clearance can be achieved without "dragging" brakes. *(Courtesy of Ford Motor Co. of Canada Ltd.)*

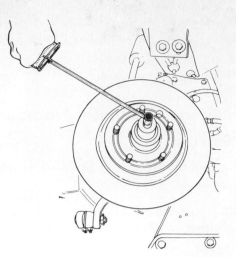

Figure 3-126. Using a torque wrench to adjust wheel bearings. Usual procedure is to tighten nut to specifications; then back off nut a specified amount and lock with cotter pin.

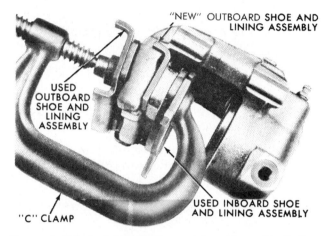

Figure 3-127. Installing brake pad in caliper with C-clamp. Antirattle materials and clips must be installed properly with pads.

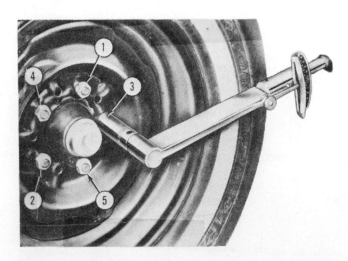

Figure 3-128. Tightening wheel nuts must be done to specified torque and in proper sequence to avoid drum or disc distortion.

justment. If no pedal free play is provided, the master cylinder piston will keep the compensating port covered and brakes will not release, since hydraulic system pressure cannot drop. Bleed the master cylinder and connect the brake lines.

Bleeding Hydraulic System

Bleeding the brake hydraulic system is necessary after reconditioning or if air has entered the system. Air in the system results in spongy pedal feel and action; since air in the system is compressible, reduced pressure and force result in poor braking action when the brakes are applied.

Two methods of bleeding brakes are normally used: manual bleeding and pressure bleeding.

To bleed the brakes manually, proceed as follows.

- Close all bleeder screws.
- Fill the master cylinder with clean brake fluid.
- Use a bleeder drain tube and jar to drain fluid from each wheel.
- Have an assistant push down the brake pedal.
- Attach the bleeder drain to the right rear-wheel bleeder.
- While the pedal is down, open the bleeder screw about a three-quarter turn.
- Close the bleeder screw when the flow stops.
- Release the pedal.
- Repeat this procedure until all air has been removed and a stream of fluid with no bubbles flows when the bleeder is opened. Keep the master cylinder full.

- Repeat this procedure next at the left rear wheel, then the right front wheel, and finally the left front wheel.
- If brakes were properly adjusted and bled, there should be a firm pedal feel with no more than half the total pedal movement required (pedal reserve).
- On brakes equipped with a metering valve, the valve may have to be held open during the bleeding process.

To *pressure bleed* the hydraulic system, a pressure tank and special master cylinder cover are required. The tank should contain an adequate supply of good brake fluid and be pressurized to approximately 25 psi (172.375 kPa). A quick couple connects the tank hose to the master cylinder cover. With all the bleeders closed, the pressure tank valve is opened, pressurizing the brake system and keeping the master cylinder full of fluid. If brakes are equipped with a metering valve, the valve must be held open during bleed-

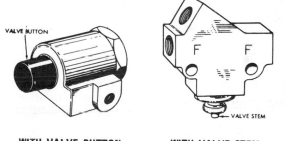

WITH VALVE BUTTON
Type "A" — Push-in Only

WITH VALVE STEM
Type "A" — Push-in
Type "B" — Pull-out

Type "A", push-in: 12 - 25 pounds spring load.

Type "B", pull-out: 22 - 35 pounds spring pull, 0.060 inch minimum travel.

Figure 3-130. Metering valve must be held open during bleeding of brakes.

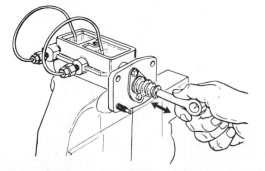

Figure 3-129. Bench bleeding of master cylinder is done as illustrated.

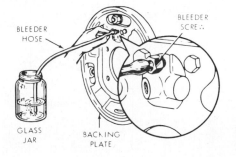

Figure 3-131. Bleeder hose and jar used for bleeding brakes.

ing. No pedal operation is required. Bleeding is then accomplished by opening the bleeder screw at each wheel in the same manner and sequence as in manual bleeding until all air is removed. Then the pressure tank valve is closed and the pressure bleeder disconnected. Replace the master cylinder cover and check pedal feel and reserve.

Power Brake Service Procedure

Vacuum Power Brake

Most automotive shops no longer overhaul vacuum power boosters. The general practice is to determine the booster's condition by performing a functional test; if the booster fails this test; it is replaced with a new or rebuilt unit.

Functional Test

1. With the engine stopped, depress the brake pedal three or four times to eliminate all vacuum in the booster.

2. Hold the brake pedal down and start the engine. When the engine starts, the brake pedal should move down slightly, indicating power assist.

3. Release the brake pedal. Stop the engine and let it stand for 15 to 30 minutes. After this period of time there should be sufficient vacuum in the unit for several power-assisted brake applications. If no vacuum is present at this time, there is a leak in the vacuum booster check valve.

4. Loss of brake fluid from the master cylinder reservoir can sometimes be traced to a fluid leak from the master cylinder into the booster. Such a leak is not visible externally but may be detected by traces of brake fluid in the vacuum booster line to the intake manifold. This requires master cylinder repairs and cleaning or replacement of the booster.

Hydro-Boost Power Brake

Hydro-Boost units are generally not repaired in automotive service shops. The malfunction must be diagnosed to determine whether the booster unit itself is defective or if the problem is in the pressure supply system from the power steering unit. After proper diagnosis, if the Hydro-Boost unit is found to be faulty, it must be replaced. Follow the procedures outlined in the diagnosis guide.

Final Inspection and Testing Procedure

After the brake system has been overhauled, a final inspection should be performed as follows.

1. Depress the brake pedal firmly to pressurize the system. Check for hydraulic fluid leaks while the system is pressurized.

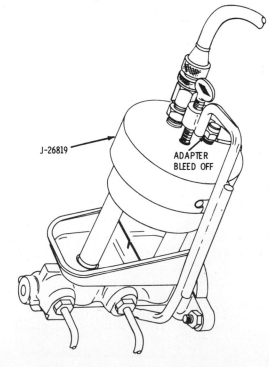

Figure 3-133. Typical master cylinder pressure bleeder adapter for plastic reservoir master cylinder. *(Courtesy of General Motors Corporation)*

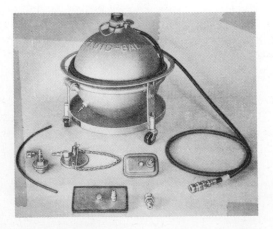

Figure 3-132. Pressure bleeding equipment with special covers for different types of master cylinders.

2. The pedal should remain firm and not move down while constant pressure is applied. There should be adequate pedal reserve.

3. Check the parking brake operation. Brakes should hold firmly when the parking brake is applied and release completely when released.

4. The brake warning light should be on when the parking brake is applied but should not be on when service brakes are applied.

5. Rear brake lights should go on when brakes are applied.

6. Road test the vehicle to determine whether brakes operate properly. There should be no pull to either side when braking; no excessive front-end dive or rear-wheel lock-up; no squeal, rattle, or chatter from the brakes; and no brake drag (brakes not fully releasing).

7. As with any service job, customer acceptance of a vehicle is very important. It does not matter how good a job has been done on the vehicle if grease, oil, and smudges appear on the windows, doors, steering wheel, and upholstery; the customer will not be happy. A job well done leaves the car with no such evidence that it has been serviced.

PART 8 SKID CONTROL SYSTEMS

Systems Operation

Braking is most effective while the wheels are still rotating just before a skid or wheel lock-up takes place. A skid control system is designed to prevent skid or wheel lock-up. The system consists of a speed sensor at each wheel, which sends an electric signal to a control modulator, which reduces hydraulic pressure to the wheels, which signal an impending skid. The system does not affect normal brake operation except during an impending skid.

PART 9 SKID CONTROL SYSTEM DIAGNOSIS AND SERVICE PROCEDURE

Servicing the skid control system varies considerably from one model to another and

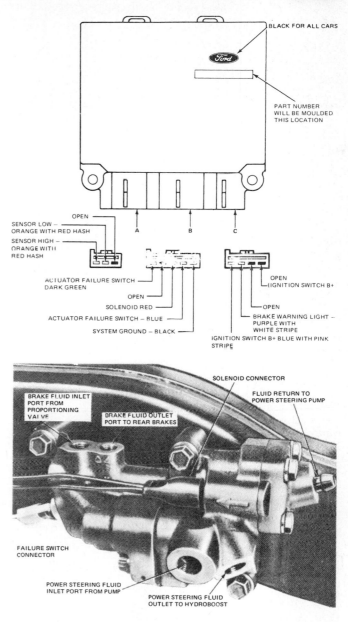

Figure 3-134. Control module multiple connectors (above) and actuator assembly (below) for Ford Sure-Track Brake System. *(Courtesy of Ford Motor Co. of Canada Ltd.)*

should be done according to shop manual prodecures. Test include system functional tests, electrical control system tests, sensor tests, solenoid tests, system ground tests, and failure switch tests.

The following diagnostic chart is an example of diagnostic procedures on a late model Ford system. Service consists of correcting electrical system faults and replacing faulty components.

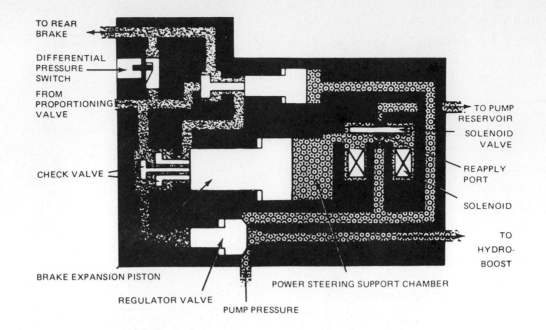

Schematic—Actuator In Normal Position

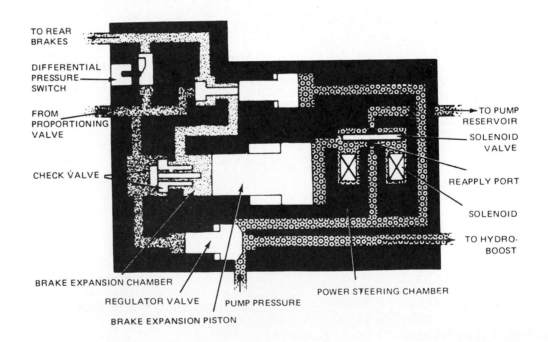

Schematic—Actuator In Activated Position

Figure 3-135. Actuator operation in normal position (above) and in activated position (below). *(Courtesy of Ford Motor Co. of Canada Ltd.)*

DIAGNOSIS CHART: SURE-TRACK BRAKE SYSTEM

PROBLEM	CAUSE	CORRECTION
Brake warning light comes on immediately and stays on after key is turned to run position.	1. Check sure-track 4-amp fuse. 2. Differential valve shuttled. 3. Short in brake warning light ground circuit. 4. Open B+ lead to module. 5. Shorted B+ lead to module.	1. Replace fuse. 2. Refer to hydraulic brake system diagnostic procedure. 3. Remove plug C from module. Turn ignition to *run*. If light comes on, locate and repair short. 4. Check B+ wiring and connectors for open circuit and repair as required. 5. Check B+ wiring and connectors for short circuit and repair as required. Perform solenoid test and repair as required.
Brake warning light flashes and then comes on and stays on 4 to 6 seconds after ignition key is turned to run position.	1. Open speed sensor, circuit connections, and/or circuitry. 2. Open actuator failure switch, circuit connections, and/or circuitry. 3. Closed actuator differential failure switch in actuator. 4. Open actuator solenoid, circuit connections, and/or circuitry. 5. Incorrect computer module is installed.	1. Check sensor, harness connections at computer module, in trunk, and at sensor, and repair as required. 2. Check failure switch, connections at actuator, computer module, and in engine compartment. Repair as required (perform failure switch test). 3. Perform failure switch test. Perform hydraulic brake system diagnostic procedure. 4. Check solenoid plus connectors at computer module in passenger compartment and at actuator. Repair as required (perform solenoid test). 5. Replace black (white on Granada and Monarch) case.
Brake warning light does not flash and go out as soon as the key is turned to run position and: (a) Brake warning light comes on when the key is turned to start. (b) Brake warning light does not come on when key is turned to start.	1. Loose or missing skid control ground wire. 2. Connectors loose or not connected to skid control module. 3. Skid control module not installed. 4. Burned out brake warning light bulb. 5. Loose or broken wire in brake warning light circuit.	1. Repair ground wire (perform system ground test). 2. Replace and/or repair connectors. 3. Install skid control module. 4. Replace bulb. 5. Replace or repair wire.
Brake warning light operates normally (flashes) and skid control system cycles during rough road conditions or normal braking.	1. Loose ground connection. 2. Loose sensor connection. 3. Worn or damaged sensor or sensor rotor. 4. Loose B+ and failure light connector at computer module.	1. Perform system ground test procedure. 2. Check sensor connection at computer module in trunk, and at sensor. Repair as required. 3. Perform sensor test. 4. Check plug C and repair as required.
Brake warning light operates normally (flashes) and actuator cycles slowly or not at all during maximum braking condition.	1. Shorted sensor circuit or worn or damaged sensor or rotor. 2. Plugged actuator filter.	1. Perform sensor test. 2. Replace actuator.

If the corrections listed do not correct the observed condition, it may be necessary to replace the computer module.

PART 10 AIR BRAKES

Air brake operation is similar to hydraulic brake operation except that compressed air is used to actuate the brakes instead of hydraulic pressure. Air brakes are used on heavier trucks and equipment. Typical system components are illustrated in Figure 3-136.

When the brakes are applied by operating a foot valve or a hand-operated valve, air pressure is applied directly to the brake shoes through a diaphragm in the brake chamber through mechanical linkage.

The engine-driven compressor provides air pressure to the pressure tanks at ap-

proximately 100 to 130 psi (689.5 to 896.35 kPa). When pressure reaches approximately 130 psi (896.35 kPa), the governor cuts out the compressor operation. When the brakes are applied, pressure drops. When pressure drops to approximately 100 psi (689.5 kPa), the governor again cuts in the compressor to raise system pressure. A warning buzzer warns the driver when system pressure drops to about 60 psi (413.7 kPa).

Compressor and governor operation are illustrated in Figure 3-137. The brake-actuating mechanism is either a wedge or a cam. The cam type is shown in Figure 3-138. Brake chamber and parking brake operation are shown in Figure 3-139. Quick-release valves are located near the brake chambers to exhaust air rapidly for quick brake release.

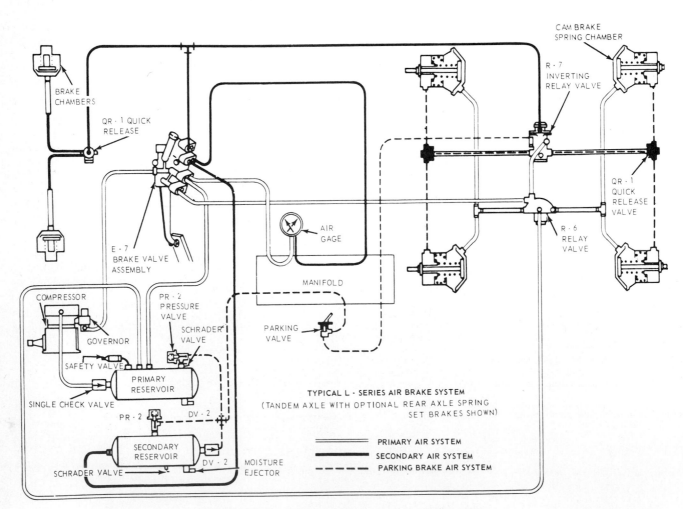

Figure 3-136. Typical air brake system components.

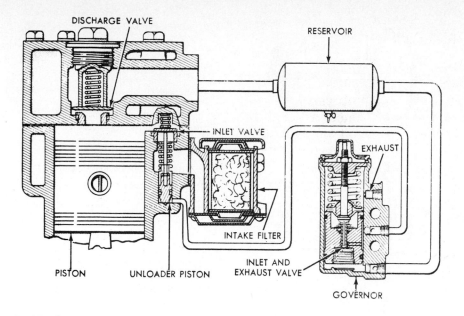

Figure 3-137. Compressor and governor operation. The engine-driven compressor produces the compressed air while the governor tells the compressor when to do it.

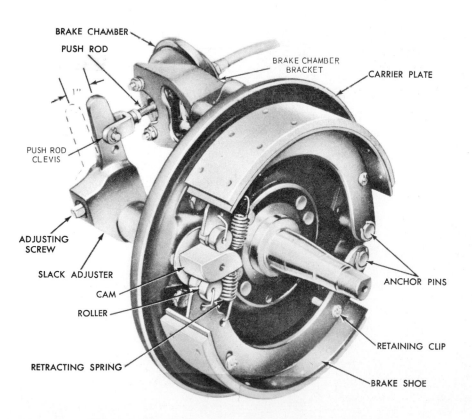

Figure 3-138. Cam-type air brake actuating mechanism.

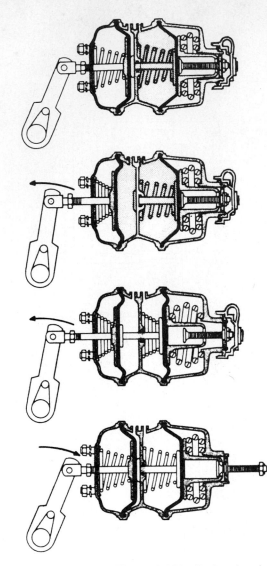

NORMAL DRIVING

A SAFE LEVEL OF AIR PRESSURE WITHIN THE SYSTEM HOLDS SPRING BRAKES RELEASED, BUT ALWAYS READY FOR PARKING OR EMERGENCY APPLICATION.

NORMAL SERVICE BRAKE

WITH THE SEPARATION OF THE TWO UNITS, THE SPRING BRAKE CANNOT INTERFERE WITH THE OPERATION OF THE NORMAL SERVICE BRAKE.

PARKING BRAKES

A FINGER-TIP CONTROL IN THE CAB TO EXHAUST AIR PRESSURE WITHIN THE SPRING BRAKE GIVES THE DRIVER FOOLPROOF AND POSITIVE PARKING BRAKES.

EMERGENCY BRAKES

THE SPRING BRAKES ARE INSTALLED TO OPERATE AUTOMATICALLY UPON LOSS OF AIR PRESSURE.

MANUAL RELEASE

THE BUILT-IN MANUAL RELEASE ALLOWS EASY RELEASE TO RELINE BRAKES OR MOVE THE VEHICLE IN THE ABSENCE OF AIR PRESSURE.

Figure 3-139. Brake chamber and parking brake operation.

AIR BRAKE SYSTEM DIAGNOSTIC CHART

TRUCKS, TRACTORS, BUSES

Insufficient Brakes
Brakes need adjusting, lubrication, or relining.
Wrong type brake lining.
Poor fit between lining and drum.
Low air pressure (below 80 psi) (551.6 kPa).
Brake valve defective; not delivering pressure.
Incorrect angle between slack adjuster and brake chamber push
    rod.

Brakes Apply Too Slowly
Brakes need adjusting or lubricating.
Low air pressure in the brake system (below 80 psi) (551.6 kPa).
Brake valve delivery pressure below normal.
Excessive leakage when brakes applied.
Restricted tubing or hose line.
Binding in camshaft or anchor pins.
Binding in brake linkage.

Brakes Release Too Slowly
Brakes need adjusting or lubricating.
Brake valve not returning to fully released position.
Restricted tubing or hose line.
Exhaust port of brake valve or quick-release valve restricted or
    plugged.
Defective brake valve or quick-release valve
Binding in camshaft or anchor pins.
Binding in brake linkage.

Brakes Grab
Grease on brake lining; reline brakes.
Brake drum out of round.
Defective brake valve.
Brake rigging binding.
Wrong type brake lining.

Uneven Brakes
Brakes need adjusting, lubricating, or relining.
Grease on lining.
Brake shoe return spring or brake chamber spring weak or
    broken.
Brake drum out of round.
Leaking brake chamber diaphragm.

TRAILERS

Insufficient Brakes
Same as for trucks except may also be caused by defective relay-
    emergency valve.
Restricted tubing (service line).

Brakes Apply Too Slowly
Same as for trucks.
Excessive air leakage with brakes applied.

Brakes Release Too Slowly
Same as for trucks.
Exhaust port of relay-emergency valve restricted or plugged.

Brakes Do Not Apply
Brake system not properly connected to brake system of tractor.
Tractor protection valve malfunctioning.
No air pressure.
Plugged tubing or hose.

Brakes Do Not Release
Brake system not properly connected to towing vehicle.
Relay-emergency valve in emergency position.
Tractor protection valve malfunctioning.

Brakes Grab
Same as for trucks.
Defective relay emergency valve.

Uneven Brakes
Same as for trucks.

PART 12 ELECTRIC BRAKES

Electric brakes are most commonly used on trailers. The friction components are basically the same as for hydraulic brakes; however, instead of using a hydraulic system to actuate the brakes, an electrical system is used. A foot- or hand-operated rheostat is used to control the amount of electrical current to an electromagnet. The electromagnet operates a cam, which in turn pushes the shoes against the drum. When current stops, the cam and brake shoes are returned to the release position by springs. The severity of braking increases with increased current.

A variation of the hand- or foot-operated rheostat is one sensitive to brake hydraulic system pressure . As pedal pressure is incresed, hydraulic pressure actuates the rheostat to increase current to the electromagnet. Another type uses an inertia weight to uncover a light. As the vehicle slows, the weight or pendulum in the control device moves forward, exposing a light-sensitive diode to more of the light beam and thereby increasing current to the electromagnet in the trailer brake.

PART 13 ELECTRIC BRAKE SYSTEM DIAGNOSIS AND SERVICE PROCEDURE

The shoes, linings, drums, and attaching parts are serviced the same as in hydraulic brake systems. As in any electrical system, all connections must be clean and tight between the control device and the brake units of the wheels. This includes the ground circuit. Poor connections or grounds resist current flow and reduce braking effectiveness. If all connections are clean and tight and circuit continuity has been established, the rheostat and magnet must be checked for proper operation. If either of these units is at fault, it must be replaced.

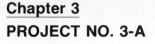

Chapter 3

PROJECT NO. 3-A

BRAKES

TITLE: <u>HYDRAULIC SYSTEM</u>

STUDENT'S NAME _____

PERFORMANCE OBJECTIVES

After sufficient opportunity to study this portion of the text and the appropriate training models and with the instructor's supervision and demonstrations, you should be able to perform the following tasks at the request of your instructor.

TASK 1---Describe the basic principles of the brake hydraulic system including the following:
- Transmitting force and motion
- Calculating pressures, force, and motion.

TASK 2---Define *friction.* Calculate the coefficient of friction as indicated on the diagram. Describe the effects of vehicle weight and speed increase on friction and heat.

TASK 3---Name the components indicated on the hydraulic system diagram, the master cylinder diagram, the wheel cylinder diagram, and the caliper diagram.

TASK 4---State the purpose of the hydraulic brake system; describe its construction and operation under normal and abnormal operating conditions. Describe the long-term effects of such operation on the system and its components.

TASK 5---State the purpose of the parking brake system; describe its construction and operation under normal and abnormal conditions. Name the parts indicated on the parking brake diagram.

PERFORMANCE EVALUATION

Your instructor may require you to perform these tasks in any of the following ways in order to evaluate your performance:
- By asking test questions
- By asking you to describe the performance of these tasks in writing
- By asking you to describe the performance of these tasks orally

BRAKES

TITLE: <u>POWER BRAKES</u>

STUDENT'S NAME _____

PERFORMANCE OBJECTIVES

After sufficient opportunity to study this portion of the text and the appropriate training models and with the instructor's supervision and demonstrations, you should be able to perform the following tasks at the request of your instructor.

TASK 1---State the purpose of the vacuum power brake booster. Describe its construction and operation. Name the components indicated on the vacuum booster diagram.

TASK 2---State the purpose of the Hydro-Boost power brake unit. Describe its construction and operation. Name the components indicated on the Hydro-Boost diagram.

PERFORMANCE EVALUATION

Your instructor may require you to perform these tasks in any of the following ways in order to evaluate your performance:
- By asking test questions
- By asking you to describe the performance of these tasks in writing
- By asking you to describe the performance of these tasks orally

BRAKES

TITLE: <u>BRAKE SYSTEM</u>
<u>DIAGNOSIS</u>

STUDENT'S NAME _____

PERFORMANCE OBJECTIVES

After sufficient opportunity to study this portion of the text and the appropriate training models and with the instructor's supervision and demonstrations, you should be able to perform the following tasks at the request of your instructor.

TASK 1---Complete the diagnostic chart provided.

TASK 2---Diagnose brake problems on a vehicle specified by your instructor.

PERFORMANCE EVALUATION

Your instructor may require you to perform these tasks in any of the following ways in order to evaluate your performance:

- By asking test questions
- By asking you to describe the performance of these tasks in writing
- By asking you to describe the performance of these tasks orally

PROBLEM	CAUSE	CORRECTION
Pedal bottoms - no brakes.	1. 2. 3. 4. 5.	1. 2. 3. 4. 5.
Pedal soft and spongy.	1. 2. 3. 4.	1. 2. 3. 4.
Pedal hard, little braking.	1. 2. 3. 4. 5. 6. 7.	1. 2. 3. 4. 5. 6. 7.
Pedal pulsates.	1. 2. 3. 4.	1. 2. 3. 4.
Brakes grab.	1. 2. 3. 4. 5.	1. 2. 3. 4. 5.

PROBLEM	CAUSE	CORRECTION
Car pulls to one side when braking.	1. 2. 3. 4.	1. 2. 3. 4.
Brakes drag.	1. 2. 3. 4.	1. 2. 3. 4.
Brakes chatter.	1. 2. 3. 4. 5. 6.	1. 2. 3. 4. 5. 6.
Brakes squeal.	1. 2. 3. 4. 5. 6. 7. 8.	1. 2. 3. 4. 5. 6. 7. 8.

TITLE: BRAKE SYSTEM
SERVICE

STUDENT'S NAME _____

PERFORMANCE OBJECTIVES

After sufficient opportunity to study this portion of the text and the appropriate training models and with the instructor's supervision and demonstrations, you should be able to perform the following tasks at the request of your instructor.

TASK 1--- Overhaul a disc-drum brake system including the hydraulic system according to procedures and specifications given in the text, the appropriate shop manual, and by your instructor. Record the results on the worksheet provided.

PERFORMANCE EVALUATION

Your ability to perform these tasks will be evaluated by your instructor on the basis of how accurately you have followed the procedures and specifications given to you by your instructor, by the text, and by the manufacturer's service manual.

— MASTER CYLINDER   OK _____   NOT OK _____   EXPLAIN _____

— FLUID LEVEL   OK _____   NOT OK _____   EXPLAIN _____

— FLUID CONDITION   OK _____   NOT OK _____   EXPLAIN _____

— BRAKE PEDAL HEIGHT   OK _____   NOT OK _____   EXPLAIN _____

   FREE PLAY   OK _____   NOT OK _____   EXPLAIN _____

— STOP LIGHTS   OK _____   NOT OK _____   EXPLAIN _____

— BRAKE WARNING LIGHT   OK _____   NOT OK _____   EXPLAIN _____

— POWER BRAKE UNIT   OK _____   NOT OK _____   EXPLAIN _____

— BRAKE LINES   OK _____   NOT OK _____   EXPLAIN _____

   STEEL LINES   OK _____   NOT OK _____   EXPLAIN _____

   FLEX LINES   OK _____   NOT OK _____   EXPLAIN _____

   RETAINING CLIPS   OK _____   NOT OK _____   EXPLAIN _____

— LEAKS   OK _____   NOT OK _____   EXPLAIN _____

— BRAKE DRUMS OR DISCS

   LEFT FRONT   OK _____   NOT OK _____   EXPLAIN _____

   RIGHT FRONT   OK _____   NOT OK _____   EXPLAIN _____

MAXIMUM ALLOWABLE DIMENSION _____

ACTUAL AFTER MACHINING   L.F. _____   R.F. _____

LEFT REAR   OK _____   NOT OK _____   EXPLAIN _____

_____

RIGHT REAR   OK _____   NOT OK _____   EXPLAIN _____

_____

MAXIMUM ALLOWABLE DIMENSION _____

ACTUAL AFTER MACHINING   L.R. _____   R.R. _____

— BRAKE LINING

   FRONT   OK _____   NOT OK _____   EXPLAIN _____

   REAR   OK _____   NOT OK _____   EXPLAIN _____

— RETRACTING SPRINGS   OK _____   NOT OK _____   EXPLAIN _____

— HOLD DOWN SPRINGS   OK _____   NOT OK _____   EXPLAIN _____

— ADJUSTING SCREW   OK _____   NOT OK _____   EXPLAIN _____

— SELF-ADJUSTING   OK _____   NOT OK _____   EXPLAIN _____
  CABLES, LINKS AND
  LEVERS

— PARK BRAKE CABLE   OK _____   NOT OK _____   EXPLAIN _____

— PARK BRAKE STRUT,
  SPRING AND LEVER   OK _____   NOT OK _____   EXPLAIN _____

— BACKING PLATES   OK _____   NOT OK _____   EXPLAIN _____

— PARK BRAKE PEDAL OR
  LEVER MECHANISM   OK _____   NOT OK _____   EXPLAIN _____

— WHEEL CYLINDERS
  LEFT FRONT   OK _____   NOT OK _____   EXPLAIN _____
  RIGHT FRONT   OK _____   NOT OK _____   EXPLAIN _____
  LEFT REAR   OK _____   NOT OK _____   EXPLAIN _____
  RIGHT REAR   OK _____   NOT OK _____   EXPLAIN _____

— AXLE AND WHEEL SEALS OK _____ NOT OK _____ EXPLAIN _____

— WHEEL BEARINGS
  REPACKED   OK _____   NOT OK _____   EXPLAIN _____

— WHEEL BEARING
  ADJUSTMENT
  PROCEDURE AND
  SPECIFICATION

_____

— WHEEL NUT TORQUE _____

— MASTER CYLINDER   OK _____   NOT OK _____   EXPLAIN _____

— BRAKE BLEEDING   OK _____   NOT OK _____   EXPLAIN _____

— PRESSURE
  DIFFERENTIAL SWITCH OK _____ NOT OK _____   EXPLAIN _____

— METERING VALVE   OK _____   NOT OK _____   EXPLAIN _____

— BRAKE WARNING LIGHT OK _____ NOT OK _____   EXPLAIN _____

— PARKING BRAKE
  OPERATION   OK _____   NOT OK _____   EXPLAIN _____

— BRAKE PEDAL
  DEPRESSOR TEST   OK _____ . NOT OK _____   EXPLAIN _____

STUDENT SIGN HERE _____

INSTRUCTOR VERIFY _____

1. Brake system service should not be attempted without a thorough knowledge of proper repair procedures. True or false?

2. The lives of drivers, passengers, and pedestrians are literally dependent on the technician's ability to do a good brake job. True or false?

3. Wheel brake units are of either the_____ type or the_____type.

4. Two types of power brakes in use are the_____type and the_____type.

5. Electric brakes are most commonly used on_____.

6. Heavier trucks and highway transports are normally equipped with_____brakes.

7. Hydraulic brakes operate on the principle that liquids are not_____.

8. Force and motion can be transmitted through a hydraulic system, which can also multiply force through varying cylinder sizes. True or false?

9. Single-piston caliper braking action can be compared to that of a_____.

10. Rotor thickness should not vary more than_____inch.

11. Rotor run-out must be checked with a_____.

12. When the brake pedal is released, the caliper piston and brake pad are returned to the release position by the_____.

13. Self-adjusting on disc brakes takes place with the brakes being applied only when the car is moving in reverse. True or false?

14. Disc brakes' rotor parallelism must be checked with_____.

15. During brake application, the power brake unit has (a) vacuum on one side and atmospheric pressure on the other side of the diaphragm, (b) vacuum on both sides of the diaphragm, or (c) atmospheric pressure on both sides of the diaphragm.

16. Power brakes that are hard to apply with the engine running could have which of the following:

   a. A leaking master cylinder

   b. A leaking wheel cylinder

   c. A vacuum problem

   d. Air in hydraulic system

17. Air flow into the vacuum power brake unit accomplishes which of the following:

   a. Releases the brakes

   b. Applies the brakes

   c. Keeps brakes in hold position

18. In vacuum power brakes, the pressures on both sides of the control valve are which of the following:

   a. Equal at all times

   b. Equal in the hold position

   c. Equal in the apply position

   d. Equal in the release position

19. Hydro-Boost power brake units use_____ pressure from the_____pump.

20. Hydro-Boost power brakes allow for several power brake applications due to_____pressure being stored in the _____.

21. What method is used to compensate for brake pull when one half of a diagonally split brake system fails? _____

22. What is the purpose of the "quick take-up" feature of a master cylinder?_____

Performance Evaluation

After you have thoroughly studied the chapter on brakes and have had sufficient practice work on the various brake system components, you should, with the aid of a shop manual and the proper tools and equipment, be able to do the following.

1. Follow the accepted general precautions while servicing brakes.

2. Correctly disassemble all brake system components.

3. Properly clean all brake system components as recommended.

4. Accurately inspect and measure all brake system components to determine their serviceability.

5. Machine and recondition all brake system components accurately to the manufacturer's specifications.

6. Properly assemble and correctly adjust all brake system components.

7. Successfully bleed all air from a brake hydraulic system.

8. Perform the necessary inspection and testing procedures to determine the success of the brake system overhaul.

9. Prepare the vehicle for customer acceptance.

10. Complete the Self-Check with at least 80 percent accuracy.

11. Complete all practical work with 100 percent accuracy.

Chapter 4

Suspension Systems

The suspension system is designed to provide the best combination of ride quality, directional control, ease of handling, safety, stability, and service life.

PART 1 FRAMES, SPRINGS, AND SHOCK ABSORBERS

Frames

The frame is the foundation upon which the entire vehicle is built. All other vehicle components are directly or indirectly attached to the frame. Holes and brackets are provided in the frame for this purpose.

Several types of frame construction are used by different vehicle manufacturers as illustrated. Frame side rails and crossmembers can have tubular, U-channel, or boxed cross-section construction. Unibody construction has short frame sections or no separate frame at all.

Bumpers

Bumpers are normally attached to the vehicle at the front and at the rear. Energy-absorbing bumpers of the hydraulic or energy-absorbing-metal type are used.

Figure 4-1. Perimeter-type frame assembly. *(Courtesy of General Motors Corporation)*

Figure 4-2. X-type frame. *(Courtesy of General Motors Corporation)*

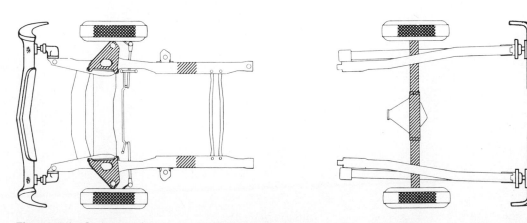

Figure 4-3. Stub-type frame as used with unibody construction. Note energy absorbing bumper mounting. *(Courtesy of General Motors Corporation)*

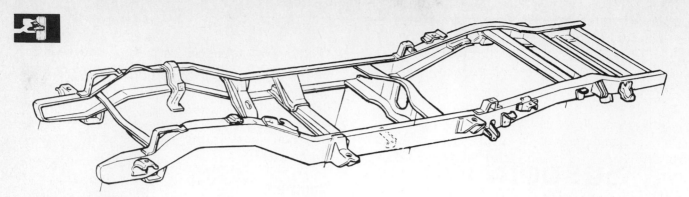

Figure 4-4. Ladder-type frame has two nearly parallel side rails with a number of crossmembers. *(Courtesy of Ford Motor Co. of Canada Ltd.)*

Springs

Several types of springs are used in suspension systems as illustrated. Springs are needed to absorb the shock of surface irregularities on the road. Tires and shock absorbers help the springs to do this job. All the weight supported by the springs is known as *sprung weight;* the weight of those components not supported by the springs is known as *unsprung weight.* The lower the proportion of unsprung weight in a vehicle, the better is the ride in general. Parts included in unsprung weight are wheels, tires, rear axle (but not always the differential), steering linkage, and some suspension parts.

Front and rear differentials on four-wheel-drive vehicles are sprung weight on some models and unsprung weight on others. Two-wheel-drive, front-wheel-drive vehicles use a transaxle in which the transmission and differential are sprung weight.

Springs are classified by the amount of deflection under a given load. This is known as the *spring rate.* Hooke's law says that a *force applied to a spring will cause a spring to compress in direct proportion to the force applied.* When that force is removed, the spring returns to its original position if not overloaded.

Heavier vehicles require stiffer springs than do lighter vehicles. Springs are designed to carry the load of the vehicle adequately

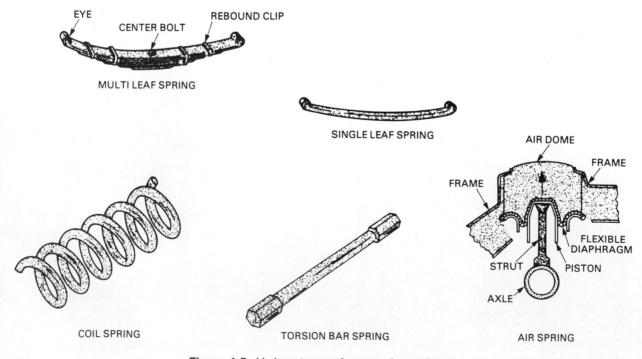

Figure 4-5. Various types of automotive springs.

while still providing as soft or smooth a ride as possible.

Coil springs, leaf springs (both multiple and single leaf), and torsion bars are the most commonly used springs. Air-type springs are used on some vehicles. Springs are rubber-mounted to reduce noise and road shock.

Leaf springs have a spring eye at one end to attach to the frame and either a slipper mount or spring eye and shackle mount at the other end. The slipper or shackles are needed because the length of the spring varies as the spring is compressed and relaxed.

Single-leaf springs are usually of the tapered plate type with a heavy or thick center section tapering off to both ends. This provides a variable spring rate for a smooth ride and good load-carrying ability.

The multiple-leaf spring has a main leaf with spring eyes at each end and a number of successively shorter leaves held together with a center bolt. The assembly is mounted to the axle housing with U-bolts, brackets, and rubber pads. The center bolt head fits in a hole in the mounting bracket of the axle housing. This prevents fore-and-aft movement of the axle, keeping it in proper alignment. Friction pads are often used between leaves to reduce friction, wear, and noise. Rebound clips are required to keep spring leaves in alignment with each other and to prevent leaf separation during rebound.

Shock Absorbers

Without shock absorbers, the continuing jounce (compression) and rebound (extension) of the spring are uncontrolled. This would be hard on the steering and suspension system, as well as provide a rough and unstable riding vehicle. The shock absorbers reduce the number and severity of these spring oscillations. Faulty shock absorbers can cause spring breakage.

Shock absorbers are mounted between the frame and the suspension by means of brackets and rubber bushings. The most common type of shock absorber is the direct acting, telescopic, hydraulic, double-acting type. This type of shock absorber is used with all types of suspension sytems.

The shock absorber operates on the principle of forcing fluid through restricted openings (orifices) on both jounce (compression) and rebound (extension). Fluid is forced from one compartment to another in the shock absorber by piston and cylinder movement (see Figures 4-8 and 4-9).

As the piston moves toward the compression head, the fluid must flow through two paths. One path is through the piston and the other is through the compression head and into the reservoir.

Fluid passes through the restriction ports,

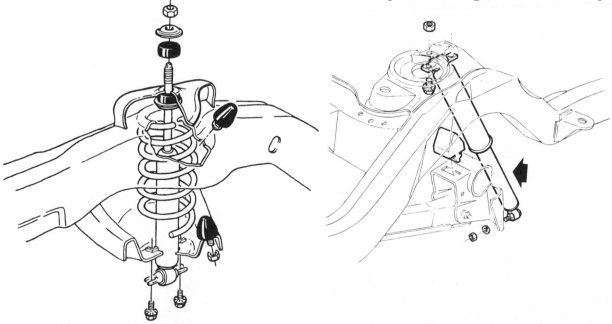

Figure 4-6. Typical front shock absorber mounting detail (left) and rear (right). *(Courtesy of Ford Motor Co. of Canada Ltd.)*

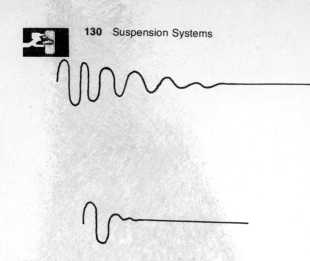

Figure 4-7. Series of spring oscillations occur as above when not controlled by shock absorber. Notice reduced number and severity of oscillations when controlled by shock absorber (below).

then under an O-ring and through the carrier orifice. Increased velocity, which produces increased pressures, lifts the O-ring assembly against the bypass spring, permitting controlled flow of the fluid.

At the same time, the fluid volume displaced by the piston rod passes through the compression head orifice. As the piston velocity increases, the fluid is forced through the compression valve restriction and then opens the valve against the valve spring. During extension, as the piston moves slowly toward the inner cylinder head, the fluid in between is forced through the restriction ports and the recoil orifice. As the piston velocity increases and more fluid is forced through these passages, pressure on the recoil valve

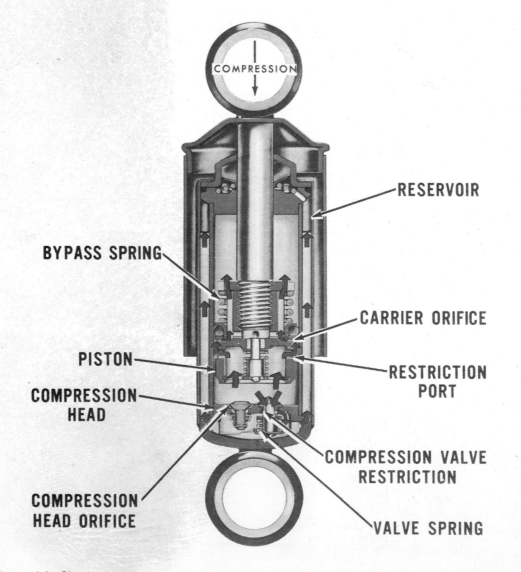

Figure 4-8. Shock absorber operation during the compression (jounce) phase. Arrows indicate direction of fluid flow through valves and restrictions. *(Courtesy of Ford Motor Co. of Canada Ltd.)*

increases until the force overcomes the valve spring, passing fluid to the opposite side of the piston. At the same time, the piston rod is wiped clean by the piston rod seal. This wiped fluid is returned to the reservoir. Fluid from the reservoir flows into the inner cylinder below the piston through the replenishing valve. This fluid volume is exactly equal to the piston rod volume that has passed through the inner cylinder head. During this entire stroke, the O-ring has been sealing the piston and inner cylinder, thus directing the fluid through the recoil control passages.

The recoil resistance of the shock absorber is controlled by the restriction ports, the recoil orifice, and the force of the recoil valve spring. Each of these may be varied independently so as to provide the damping resistance desired.

Shock absorbers are mounted vertically or at an angle. Angle mounting of shock absorbers is used to improve vehicle stability and to dampen accelerating and braking torque.

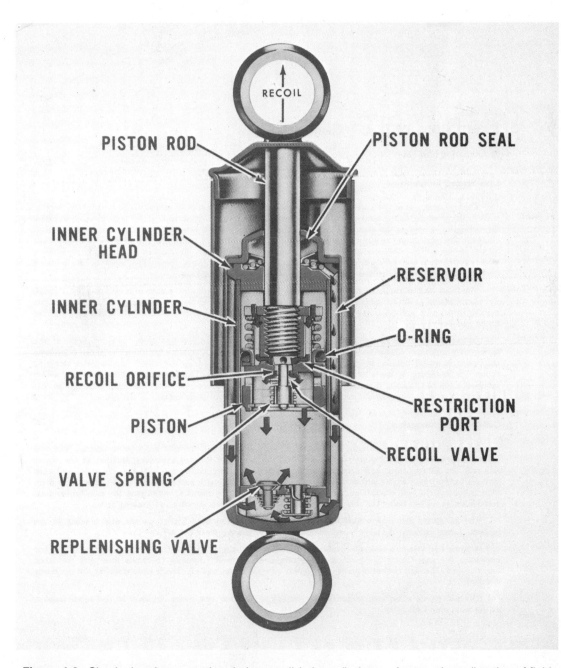

Figure 4-9. Shock absorber operation during recoil (rebound) phase. Arrows show direction of fluid flow. *(Courtesy of Ford Motor Co. of Canada Ltd.)*

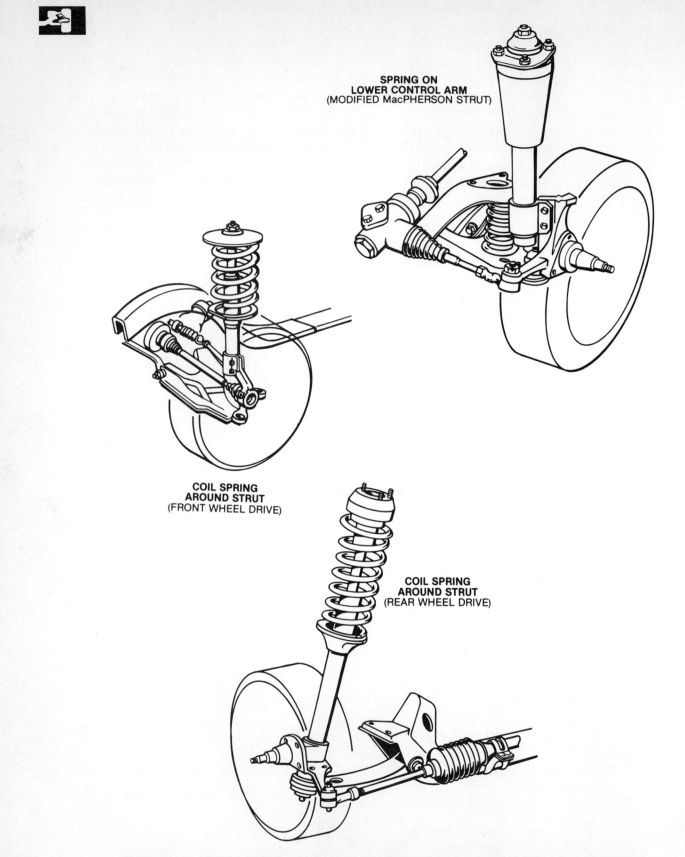

**SPRING ON
LOWER CONTROL ARM**
(MODIFIED MacPHERSON STRUT)

**COIL SPRING
AROUND STRUT**
(FRONT WHEEL DRIVE)

**COIL SPRING
AROUND STRUT**
(REAR WHEEL DRIVE)

Figure 4-10. Several designs of strut-type front suspension. *(Courtesy of Moog Automotive Inc.)*

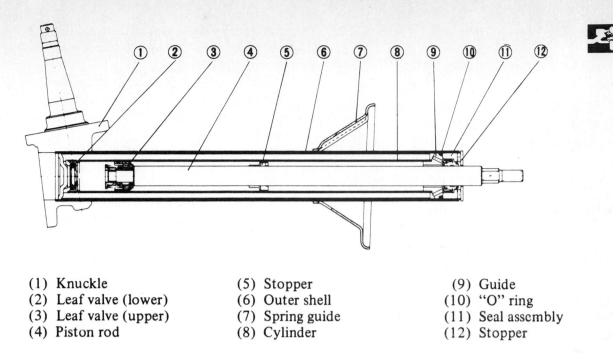

(1) Knuckle
(2) Leaf valve (lower)
(3) Leaf valve (upper)
(4) Piston rod

(5) Stopper
(6) Outer shell
(7) Spring guide
(8) Cylinder

(9) Guide
(10) "O" ring
(11) Seal assembly
(12) Stopper

Figure 4-11. Components of shock absorber-strut assembly used on front suspension of rear-wheel-drive vehicle. *(Courtesy of Chrysler Corporation)*

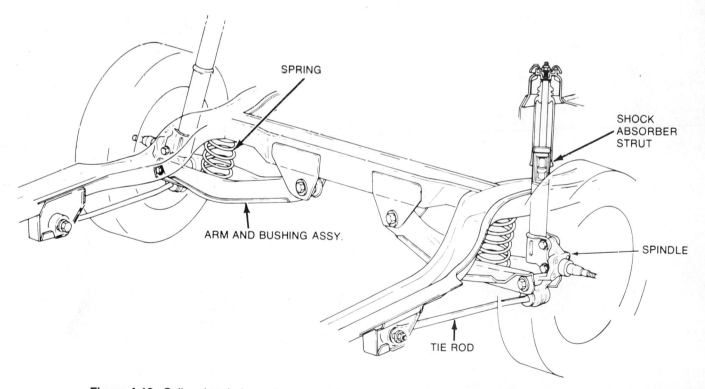

Figure 4-12. Coil-spring, independent nondriving rear suspension. *(Courtesy of Ford Motor Co. of Canada Ltd.)*

Air Shock Absorbers and Automatic Load-Leveling System

Air-adjustable rear shock absorbers are available as an option. These units help to maintain ride height when hauling heavy loads or towing trailers. Adjustment is accomplished by increasing or decreasing air pressure within the shock absorbers.

Two air shock systems are used: manually adjusted units that are inflated through air lines that connect the shocks to an air valve mounted at the rear of the automobile and an automatic load-leveling system. On the automatic system the air shocks are inflated by an on-board compressor. Adjustment is accomplished by the compressor, which is mounted in the engine compartment. The compressor is manually controlled by a three-position switch located in the compressor mounting bracket. In the automatic mode, the compressor is operated by the height sensor and relay. An auxiliary air hose may be included with the system. This hose can be connected to an auxiliary air valve on the compressor and used to inflate tires, air mattresses, and other air inflatable items. Maximum shock absorber inflation pressure is usually about 90 pounds per square inch, psi (621 kPa). Maximum allowable pressures should not be exceeded. Should the air boot leak, normal shock absorber action is not affected.

Spring Assist Shock Absorber

Spring assist shock absorbers are available for the front or rear suspension to increase the load-carrying capacity of the springs. These units consist of a conventional shock absorber with a coil spring fitted to it in a manner that causes the spring to compress when the shock absorber compresses. An upper spring seat is attached to the upper shock tube and a lower spring seat is mounted on the lower shock tube with the spring mounted in between. The spring is under some tension in its normal curb height position.

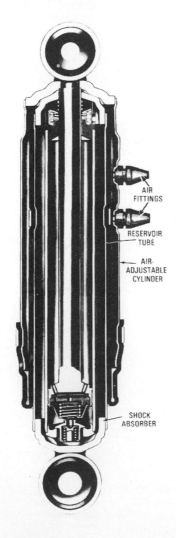

Figure 4-13. Air adjustable type of shock absorber used to maintain vehicle suspension height with increased load. *(Courtesy of General Motors Corporation)*

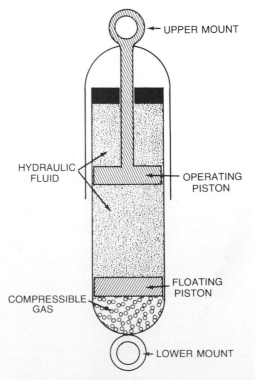

Figure 4-14. One type of gas-filled shock absorber with floating piston above gas chamber.

Gas Pressure Shock Absorber

Gas pressure shock absorbers operate on the same principle as conventional shock absorbers---with hydraulic fluid. The difference is that there is no air in the gas pressure shock, which reduces fluid aeration. Instead of a fluid reservoir, which contains some air in the conventional shock absorber, the gas pressure shock absorber has a floating piston that serves as the bottom of the lower pressure chamber. As the operating piston moves up and down, the difference in fluid volume displacement above and below the operating piston due to piston rod displacement is compensated for by the floating piston and gas chamber. The gas is either freon or nitrogen gas under pressure of 100 psi (689.5 kPa) to 360 psi (2482.2 kPa). Getting rid of fluid aeration results in more consistent shock absorber operation.

Stabilizer

The stabilizer bar (sway bar) reduces sway and stabilizes the front suspension or rear suspension. It is basically a wide U-shaped bar with one end of the U attached to each of the lower control arms through rubber mounts or links. The center section of the U is mounted to the frame at two points with rubber mounts and can pivot at these points. It is made of spring steel, which gives it the elasticity to bend or deflect and then return to its relaxed position.

When the vehicle is stationary with both wheels at the same level, there is no tension on the bar. When one wheel or the other is raised or lowered, the opposite end of the bar held by the other wheel causes the bar to twist, thereby helping to maintain the vehicle in a more level position than it would otherwise be. Thus, vehicle sway or lean in a turn is reduced.

On some vehicles the stabilizer bar is designed to act as the radius rod or strut rod as well as to provide stabilizer action. Since the ends of the stabilizer bar must be attached to the outer end of the lower control arm in any case, a relatively minor design change allows this adaptation. Naturally this reduces production costs and vehicle weight.

Some front-wheel-drive vehicles use a beam-type trailing arm rear suspension system. In this design the two wheels connected to the axle beam by means of the trailing arms

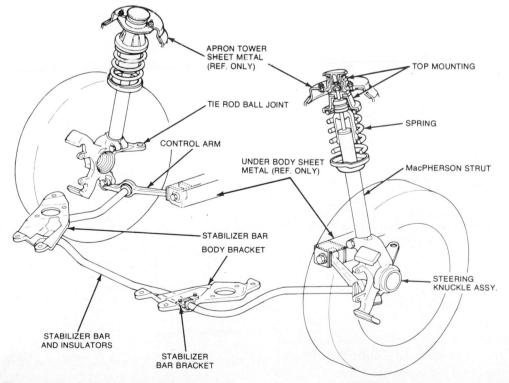

Figure 4-15. Strut-type suspension system used on many front-wheel-drive vehicles. *(Courtesy of Ford Motor Co. of Canada Ltd.)*

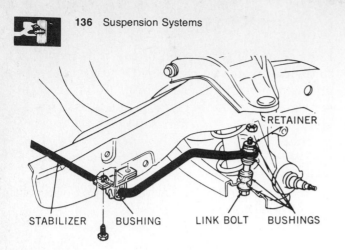

Figure 4-16. Stabilizer bar (left half shown) controls sway and stabilizes front suspension; also known as a sway bar. *(Courtesy of General Motors Corporation).*

can move up or down in a semi-independent manner. When this happens, the axle beam must twist. This provides the necessary stabilizer action as well as a degree of independent rear wheel movement.

Track Bar

The track bar is a bar usually used with coil springs; it is rubber mounted to the frame at one end and horizontally to the rear axle at the other to prevent any sideways movement between the rear axle and the body.

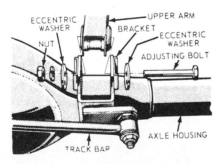

Figure 4-17. Track bar. Section shown is attached to rear axle housing, while other end is attached to car frame. The track bar keeps the body and rear axle in alignment.

PART 2 FRONT SUSPENSION

Front-suspension system design must provide good ride and stability characteristics. In addition, the front suspension must make provision for turning the front wheels to both right and left for turning corners. Straight-

ahead directional control is also provided for primarily by the front-suspension system. Because of weight transfer during braking, the front-suspension system absorbs most of the braking torque. All these factors result in the front-suspension system requiring more frequent servicing than the rear-suspension system.

Several types of front-suspension system design must be considered. These are (1) long and short arm, (2) single control arm (strut type), (3) monobeam, and (4) twin I-beam. All these suspension systems are used by manufacturers and may be found on front-wheel-drive, rear-wheel-drive, or four-wheel-drive vehicles.

Long- and Short-Arm Suspension

This independent front-suspension system has one upper control arm and one lower control arm at each front wheel. These arms are attached to the frame at the inner end of the arm through bushings that allow up and down movement of the outer ends of the arms. The outer ends of the arms are attached to a steering knuckle (spindle support) by means of ball joints. The ball joints allow the spindle to move up and down, as well as turn right and left as needed. Up-and-down movement is required to absorb road surface irregularities, while turning to right and left is required for vehicle directional control. The wheel-and-tire assembly is mounted to the spindle shaft with ball or roller bearings, a tanged washer, nut, and cotter pin.

The upper control arm is shorter than the lower arm in order to prevent the tire from scrubbing sideways during up-and-down movement of the suspension or body. This unequal

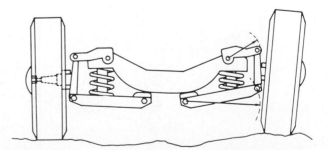

Figure 4-18. Coil-spring type independent front suspension. With independent suspension the up and down movement of one wheel does not affect the other wheel and reduces body movement as compared to solid axle suspension.

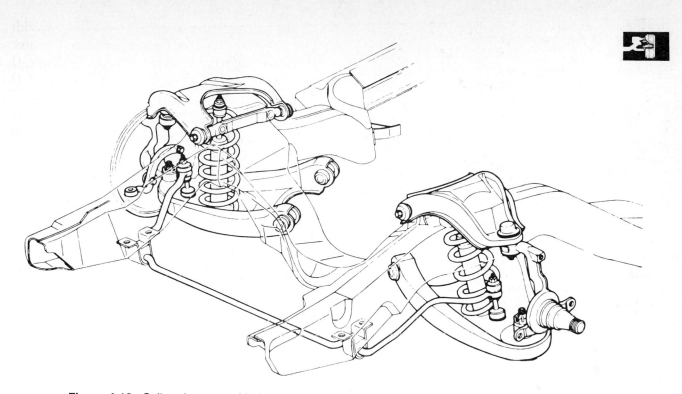

Figure 4-19. Coil-spring type of independent front suspension showing spring mounted in seat of lower control arm. Upper end of coil spring is located in a pocket in the frame. (*Courtesy of General Motors Corporation*)

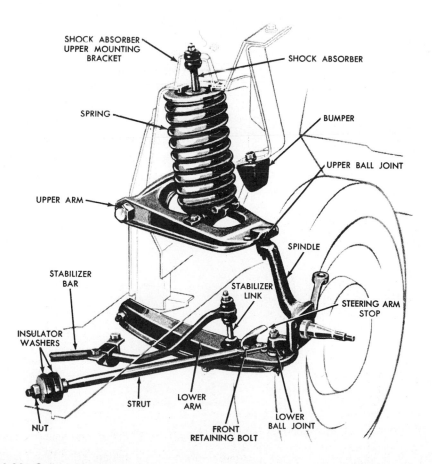

SHOCK ABSORBER
UPPER MOUNTING
BRACKET

SHOCK ABSORBER

SPRING

BUMPER

UPPER BALL JOINT

UPPER ARM

SPINDLE

STABILIZER
BAR

STABILIZER
LINK

STEERING ARM
STOP

INSULATOR
WASHERS

STRUT

LOWER
ARM

LOWER
BALL JOINT

NUT

FRONT
RETAINING BOLT

Figure 4-20. Coil spring mounted above upper control arm. Top end of spring is mounted inside a reinforced section of the body.

arm length causes the top of the wheel to move in and out with suspension movement and prevents the tire from sliding or scrubbing sideways at the bottom, where it is in contact with the road surface. Each wheel can move up and down independently, reducing the amount of body tilt as bumps on the road are encountered.

A strut rod is used with the narrow type of lower control arm to prevent fore-and-aft movement of the outer end of the control arm. The wide A-frame type of lower control arm does not need the strut rod. One end of the strut rod is mounted to the outer end of the lower control arm. The other end of the rod is rubber-mounted to the frame and pivots at this point with control arm movement.

Coil springs or torsion bars are used with independent front suspension.

Ball joints provide the pivot or hinging action for up-and-down movement and turning movement of the front wheels. One ball joint acts as the load carrier on each side while the other is the idler. The load-carrying ball joint is the lower one when the spring is mounted on the lower arm and the upper one when the spring is mounted on the upper arm.

When the load tends to pull the stud out of the ball joint, it is known as a *tension* ball joint. When the load tends to push the stud into the ball joint, it is known as a *compression* ball joint.

Ball joint studs are tapered and fit into tapered holes in the steering knuckle. A castellated nut is threaded onto the stud and locked with a cotter pin. Ball joints are provided with either a lubrication fitting or a plug that can be removed to install a fitting.

Some ball joints are provided with a wear indicator; others must be measured for wear to determine if replacement is needed. Some ball joints are riveted to the control arm; others are bolted to it, threaded into it, or are a press fit in the arm.

The spring action of leaf springs is provided by the ability of the spring to bend and then return to its original position. On torsion bars, spring action is provided by twisting the torsion bar. When a load is applied, the bar is twisted or wound in one direction; when the load is removed, the bar unwinds and returns to its original position. When a load is applied to a coil spring, it is compressed; when the load is removed, the spring extends to its original position.

Two rubber bumpers on each side (one for jounce and one for rebound) prevent metal-to-metal contact when travel limit is reached on front suspension. On rear suspension, one bumper on each side prevents metal-to-metal contact when "bottoming out."

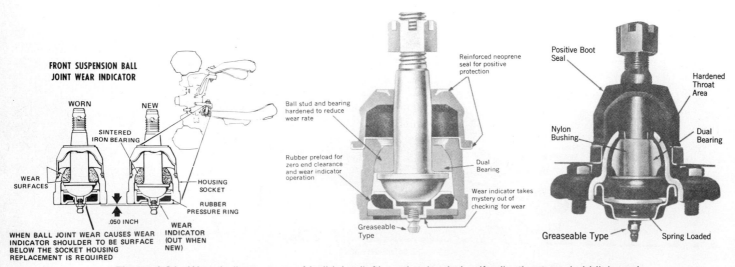

Figure 4-21. Wear indicator type of ball joint (left), spring-loaded self-adjusting type (middle), and permanently adjusted type (right).

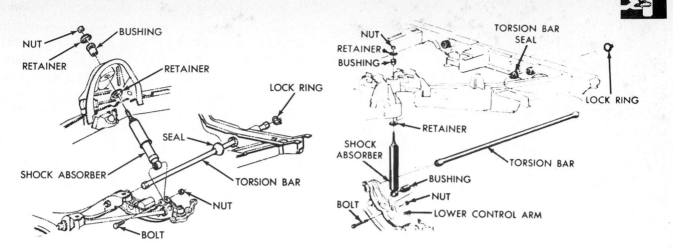

Figure 4-22. Torsion bar suspension system. Front torsion bar is mounted in hex socket at inner end of lower control arm. Rear of torsion bar is mounted in hex socket in frame crossnumber.

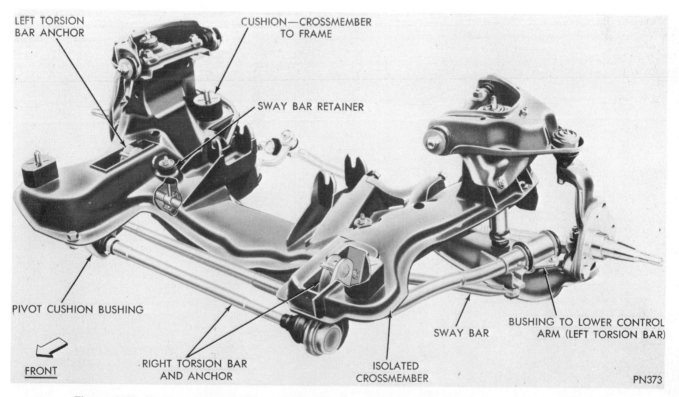

Figure 4-23. Transverse-mounted torsion bar front suspension. *(Courtesy of Chrysler Corporation)*

Single Control Arm Suspension (Strut Type)

The single control arm suspension system, also called the MacPherson strut suspension system, is based on a triangle design. The strut shaft itself is a structural member that does away with the upper control arm and bushings and the upper ball joint required on the long- and short-arm suspension. Since the strut shaft is also the shock absorber shaft, it receives a tremendous amount of force vertically and horizontally due to the accelerating and braking torque forces.

Two types of strut suspensions are used. One type has the spring surrounding the strut while the other has the spring mounted on the control arm. The former is used on both front- and rear-wheel-drive vehicles while the latter is used only on rear-wheel-drive vehicles. When the strut has the spring mounted around the strut assembly, the shock absorber, spindle, and spring are sometimes a combined unit, located at the top by the upper mount assembly, and at the bottom by the ball joint and lower control arm. The ball joint on this system is a follower ball joint. When the spring is mounted on the control arm, the ball joint is a load carrier. The lower arm may be forged or stamped steel construction. The lower arm, if it is not an A-frame design, is located either by a strut rod (compression rod) or the stabilizer bar, which can function as a combined strut rod and stabilizer bar.

The shock absorber is built into the strut outer housing. The coil spring is held in place by a lower seat welded to the strut casing and an upper seat bolted to the shock absorber piston rod. The upper mount likewise bolts to the vehicle body and is the load-carrying member through the use of a bearing or rubber bushings in most designs. In this case the coil spring and shock absorber turn right or left as the steering wheel is turned. Another design has the load-carrying bearing below the coil spring. In this case the spring does not turn when the steering wheel is turned.

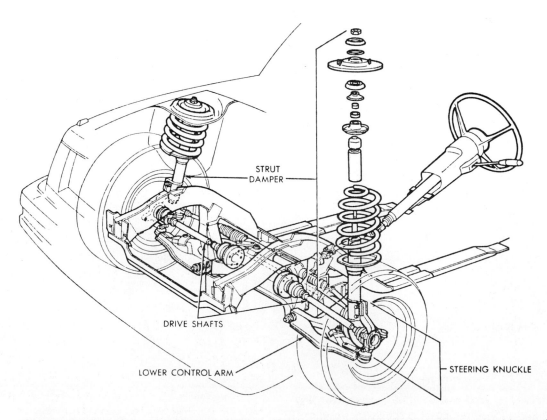

Figure 4-24. Strut type of suspension system. This system has only one control arm and is used on many compact cars. *(Courtesy of Chrysler Corporation)*

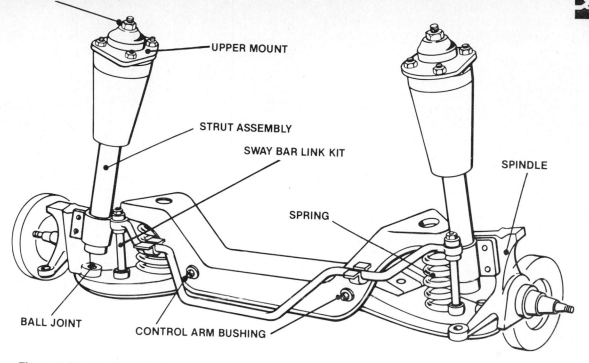

STRUT SHAFT NUT

UPPER MOUNT

STRUT ASSEMBLY

SWAY BAR LINK KIT

SPINDLE

SPRING

BALL JOINT

CONTROL ARM BUSHING

Figure 4-25. Modified strut front-suspension system used on rear-wheel-drive vehicle. *(Courtesy of Moog Automotive Inc.)*

Monobeam Suspension

The monobeam suspension is a single I-beam solid axle suspension used on trucks. It does not provide the riding comfort of independent suspension. The spindle is attached to the I-beam axle by means of a king pin (spindle bolt) and bushings. This provides for turning of the front wheels to right and left. The king pin is held in the axle by means of a lock bolt or pin. Two designs are used to attach the spindle to the axle: the Elliot and reverse Elliot designs. The Elliot design has the yoke on the axle while the reverse Elliot

has the yoke on the spindle. The reverse Elliot is easier to service when replacing king pins and bearings.

Either leaf springs or coil springs are used with this axle. In the leaf spring design, the spring assemblies maintain fore-and-aft positioning of the wheels and transfer braking torque. The coil spring design requires radius arms or rods to position the wheels fore and aft and to transfer braking torque.

Twin I-Beam Suspension

Twin I-beam suspension is a combination of the solid axle and fully independent suspension systems. Two I-beams are used with coil springs. Radius rods prevent fore-and-aft movement. On four-wheel-drive vehicles the I-beams are replaced by stamped steel axle housings, which act the same as twin I-beam suspension. One of the housings contains the differential. Either coil springs or leaf springs are used. When coil springs are used, radius rods or arms control fore-and-aft positioning of the wheels and transfer driving and braking torque. In the leaf spring design the springs serve this function.

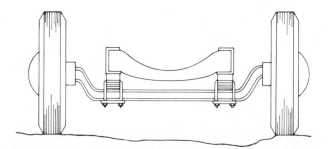

Figure 4-26. Solid axle (I-beam) front suspension with multiple-leaf springs. On this type of suspension, when one wheel rides on top of a bump, the entire vehicle is tilted much more than on independent suspension systems.

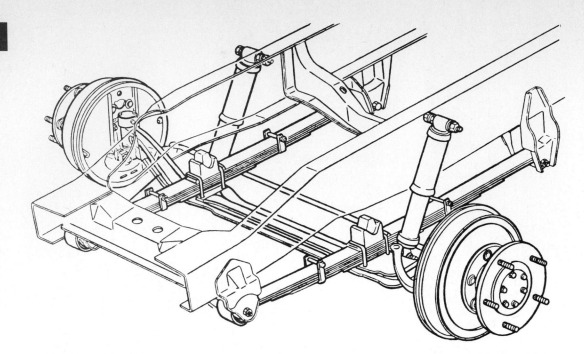

Figure 4-27. Truck-type of solid axle front suspension. *(Courtesy of General Motors Corporation)*

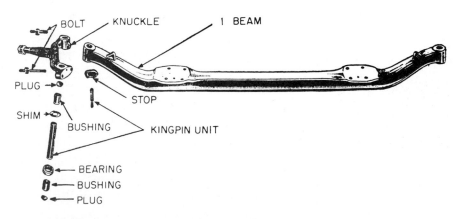

Figure 4-28. Exploded view of spindle (steering knuckle) attaching parts. When the yoke is on the spindle, as shown here, it is known as the Reverse Elliot type of axle. When the yoke is on the axle, it is called an Elliot type.

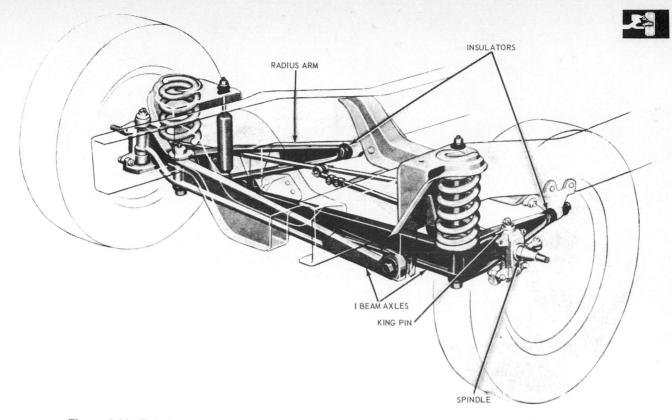

Figure 4-29. Twin I-beam type of front suspension as used on some Ford trucks. *(Courtesy of Ford Motor Co. of Canada Ltd.)*

Image labels: RADIUS ARM, INSULATORS, I BEAM AXLES, KING PIN, SPINDLE

PART 3   REAR SUSPENSION

A variety of rear-suspension system designs are being used for both driving and nondriving rear axles. This includes both nonindependent and independent suspension types.

The driving axle types must provide the means for transmitting both driving and braking torque. The nondriving rear suspension must provide the means for braking torque only. All rear-suspension systems must provide for ride quality and vehicle stability.

Nonindependent Driving Rear Suspension

Nonindependent rear-suspension types with rear-wheel drive include the coil spring and control arm type, the coil spring, trailing arm and torque arm type, and the leaf spring type.

The coil spring and control arm type transmits driving and braking torque through three or four control arms depending on design. The control arms also control fore and aft positioning of the rear axle and wheels and in some

cases lateral or side to side position of the rear axle and wheels in relation to the vehicle body. This requires two upper and two lower control arms. The lower control arms are positioned at ninety degrees to the rear axle and the upper arms at forty five degrees to the rear axle. An eccentric mount at the front of the upper control arms provides the means for adjusting the drive shaft rear universal joint operating angle. Rubber bushings at each end of the control arms provide the necessary pivoting action and also help to isolate road shock and noise transmission from the vehicle body. A track bar is used on some designs to maintain lateral alignment of the rear axle with the vehicle body. A stabilizer bar is also used on some models.

A variation of this design uses two lower trailing control arms and one long upper torque arm. The torque arm is mounted at the front to the rear of the transmission extension housing and at the rear to the differential housing.

The leaf spring non independent suspension system uses two single tapered plate or multi-leaf springs. The springs transmit driving and braking torque and provide the means for fore and aft as well as lateral positioning of the rear axle.

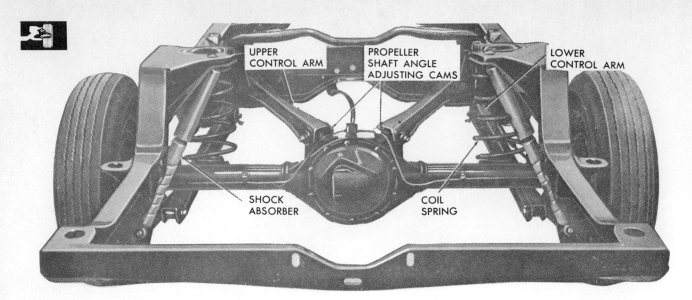

Figure 4-30. Coil-spring type of rear suspension. Two angled upper control arms control side movement. *(Courtesy of General Motors Corporation)*

Labels in figure: UPPER CONTROL ARM, PROPELLER SHAFT ANGLE ADJUSTING CAMS, LOWER CONTROL ARM, SHOCK ABSORBER, COIL SPRING

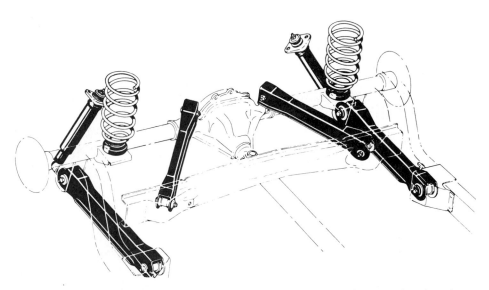

Figure 4-31. Location and mounting of control arms, shock absorbers, and springs in rear suspension.

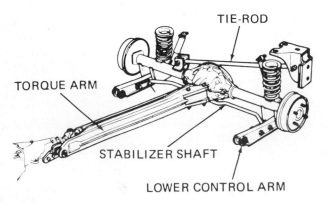

Labels in figure: TIE-ROD, TORQUE ARM, STABILIZER SHAFT, LOWER CONTROL ARM

Figure 4-32. Solid-axle, coil-spring rear suspension with a long torque arm as used on rear-wheel-drive vehicle. *(Courtesy of General Motors Corporation)*

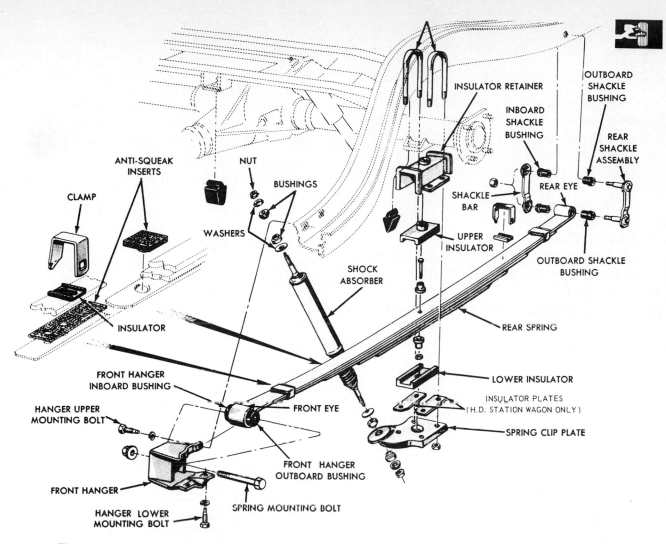

Figure 4-33. Attaching parts of rear suspension multiple-leaf spring. Note rubber bushings at front and rear of spring. *(Courtesy of Chrysler Corporation)*

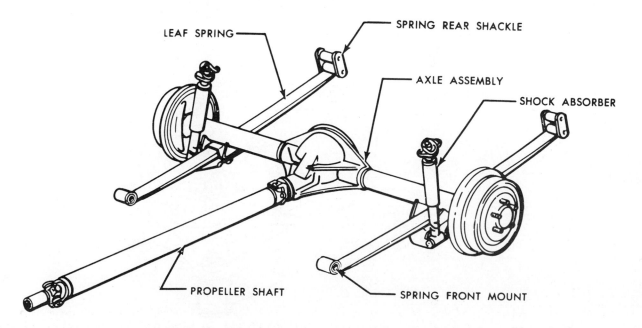

Figure 4-34. Single-leaf tapered plate spring rear suspension.

Independent Driving Rear Suspension

Several types of rear wheel drive independent suspension systems are used by different vehicle manufacturers. All of these have the differential mounted on a heavy frame crossmember. This reduces the amount of unsprung weight.

One design uses a trailing A frame type of arm and coil spring arrangement. The A frame is wide at the front where it is mounted to a heavy crossmember. This provides the means for maintaining both fore and aft and lateral positioning of the drive wheels. Rubber bushings at the front mount provide the pivoting action and help isolate the vehicle from road noise and shock. The A frame provides the lower seat for the coil spring. The upper seat is in a reinforced section of the body.

Another design uses a narrow trailing arm and a transverse multiple-leaf spring. The trailing arms control the fore-and-aft position of the wheels while two tie rods maintain lateral wheel position. Driving and braking torque are transmitted through the trailing arms. The control arms are mounted by means of rubber bushings at the front to provide pivoting action and noise isolation. The tie rods are mounted at both ends by means of rubber bushings.

A variation of the trailing arm design uses transverse torsion bar springs. The trailing arm in this case is actually a spring plate. The front end of the spring plate arm is attached to one end of a transverse torsion bar. The other end of the arm is attached to the wheel assembly. Driving and braking torque is transmitted through the arms.

A strut type of independent rear-suspension rear-wheel-drive system uses tranversely mounted A-frames and a MacPherson strut type of coil spring and shock absorber unit. Driving and braking torque is transmitted through the A-frames and struts. Rubber bushings at all mounting points provide pivoting action and noise isolation.

Nonindependent Nondriving Rear Suspension

This type of rear suspension used on some front-wheel-drive cars uses a solid axle between the two wheels, attached to the body by means of two trailing arms. The trailing arms maintain fore-and-aft positioning of the rear axle while a track bar keeps the axle in lateral alignment with the vehicle body. Rubber bushings at the front of the trailing arms and at each end of the track bar provide the pivoting action and noise isolation. Braking torque is transmitted through the trailing arms. A stabilizer bar is used to improve vehicle stability.

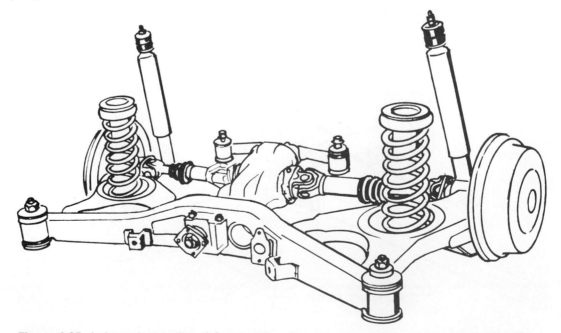

Figure 4-35. Independent trailing A-frame coil-spring rear suspension used on some rear-wheel-drive vehicles. Differential on this design is sprung weight. *(Courtesy of Moog Automotive Inc.)*

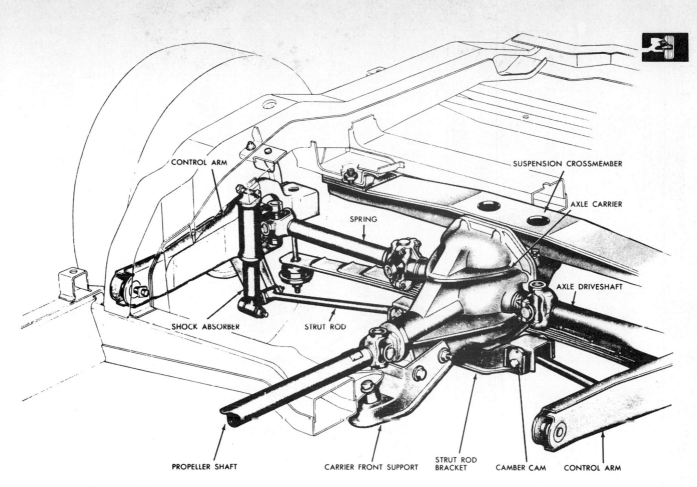

Figure 4-36. Independent rear suspension with multileaf transverse-mounted spring. *(Courtesy of General Motors Corporation)*

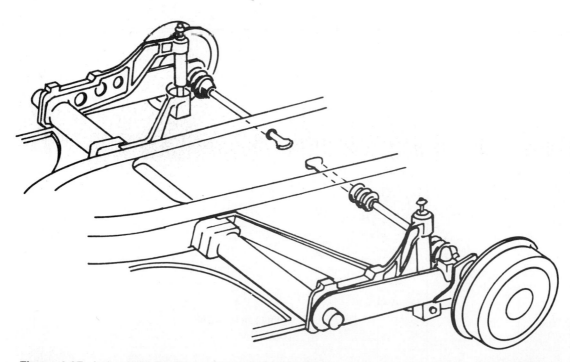

Figure 4-37. Independent trailing arm transverse torsion bar rear suspension used on some rear-wheel-drive vehicles. *(Courtesy of Moog Automotive Inc.)*

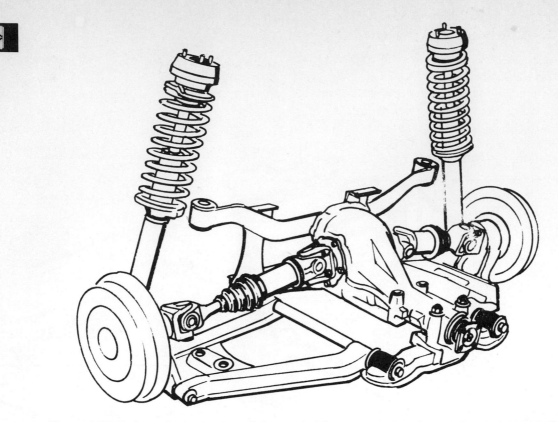

Figure 4-38. Independent transverse A-frame strut-type rear suspension used on some rear-wheel-drive vehicles. Differential is sprung weight. *(Courtesy of Moog Automotive Inc.)*

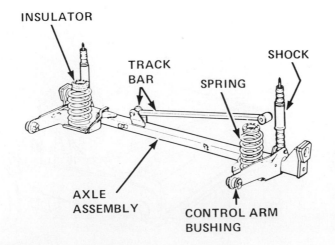

INSULATOR

TRACK BAR

SHOCK

SPRING

AXLE ASSEMBLY

CONTROL ARM BUSHING

Figure 4-39. Solid-axle, coil-spring trailing arm suspension used on some front-wheel-drive vehicles. *(Courtesy of General Motors Corporation)*

Independent Nondriving Rear Suspension

This system uses two transverse arms, coil springs, strut-shock absorbers, and tie rods. The two arms control the lateral positioning of the wheels and provide the pivoting action. The shock absorber-strut assemblies are attached at the lower end to the spindle, which is also attached to the outer end of the transverse arm. The upper end of the shock-strut is attached to a reinforced section in the body. Tie rods attached at the rear to the spindle and at the front to the torque box on the body maintain fore-and-aft wheel positioning. Braking torque is transmitted through the struts and tie rods. Rubber bushings at all mounting points provide the pivoting action and noise isolation.

Semi-Independent Nondriving Rear Suspension

Many front-wheel-drive cars use this type of rear suspension. Two trailing arms attached at the front to an axle beam are used in this system. The arms and axle assembly are attached to the vehicle by mounting brackets and rubber bushings. Up-and-down movement of either wheel imparts a twisting action to the axle beam. This provides a stabilizer effect as well as allowing semi-independent wheel movement. A coil spring and shock absorber-strut assembly are attached at the bottom to the rear of the trailing arm and at the top to a reinforced section of the body. Braking torque is transmitted through the trailing arms and struts. Pivoting action and noise isolation are provided by means of rubber bushings at all affected mounting points. Fore-and-aft and lateral positioning of the wheels is maintained by the trailing arms and the struts.

PART 4   SUSPENSION-RELATED CHARACTERISTICS

Scrub Radius

The scrub radius is the distance between the point where the projected steering axis pivot line contacts the road and the effective

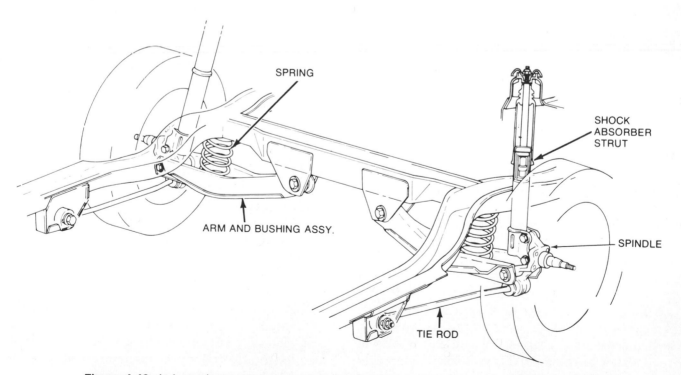

Figure 4-40. Independent transverse-arm, coil-spring rear suspension used on some front-wheel drive vehicles. *(Courtesy of Ford Motor Co. of Canada Ltd.)*

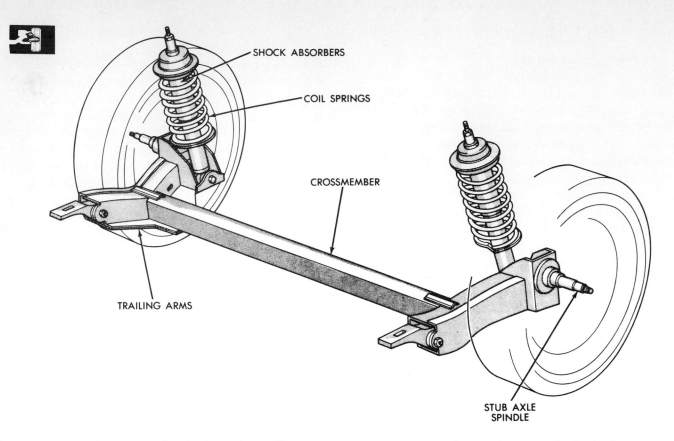

Figure 4-41. Semi-independent trailing-arm, strut-type rear suspension used on some front-wheel-drive vehicles. Axle crossmember twists to provide a measure of independent rear wheel action and also provides stabilizer bar benefits. *(Courtesy of Chrysler Corporation)*

contact point of the tire on the road. It is determined by the steering axis inclination (SAI) angle and the camber angle of the vehicle. If the contact point of the steering axis pivot line is inside the actual tire contact point, it is referred to as a positive scrub radius. If the contact point of the steering axis pivot line is outside the actual tire contact point, it is referred to as negative scrub radius. On the other hand, if the steering axis pivot line contact point and the actual tire contact point were to intersect at the road surface, this would be zero scrub radius. This is neither desirable nor practical.

Both the positive and the negative scrub radius designs provide a pivot center, which allows the tire to roll around the pivot line contact point during steering. This reduces the amount of tire squirm and distortion during steering as compared to a zero scrub radius. An excessive amount of scrub radius, however, will increase steering effort and the effects of road shock to the steering and suspension systems. A good example of increased positive scrub radius is the use of offset wheels to accommodate wider-than-standard tires.

The effect of a positive scrub radius is to tend to turn the wheel outward (toe out) while driving and during braking on non-driving front wheels. On front-wheel-drive vehicles the effect would be to tend to turn the wheel inward during driving and outward during braking. The effect of a negative scrub radius is to tend to turn the wheel inward during braking. This effect is used to offset the brake imbalance created by failure of one half of a diagonally split hydraulic system. A negative scrub radius would also tend to turn the wheel outward during torque application (driving) on front-wheel-drive vehicles.

The vehicle manufacturer chooses a scrub radius design that provides the best possible handling characteristics, ride quality, and life expectancy of tires and steering and suspension system components in concert with other design factors of the vehicle. This combination of design objectives is at best a compromise between the various factors involved. Scrub radius is affected by suspension system curb height, curb weight, camber setting, steering axis inclination, wheel and tire diameter, and wheel mounting surface to rim offset.

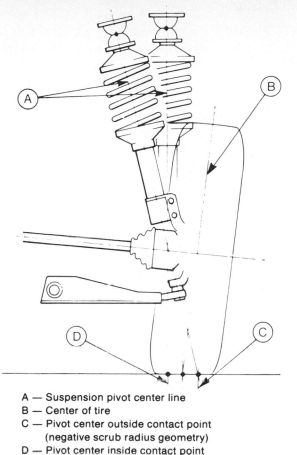

A — Suspension pivot center line
B — Center of tire
C — Pivot center outside contact point
(negative scrub radius geometry)
D — Pivot center inside contact point
(positive scrub radius geometry)

Figure 4-42. Suspension design determines whether a vehicle will have a negative or a positive scrub radius, as shown here. The scrub radius is the product of the steering axis inclination angle A and the wheel center line B where they intersect the road surface. A negative scrub radius is necessary on a front-wheel-drive vehicle with a diagonally split brake hydraulic system. *(Courtesy of Ford Motor Co. of Canada Ltd.)*

Instant Center

The angle of the upper and lower control arms in the vehicle's at-rest position (curb height and weight) determines the instant center of the vehicle suspension design. When the inner pivot points of the upper and lower control arms are closer together than the outer pivots (ball joints), the instant center is located inboard of the car. When the outer pivots are closer together than the inner pivots, the instant center is located outside the car. Some long- and short-arm suspension systems use the former design, while the latter design is used on other models.

The proportionate length of the arms and the angle of the arms determine the amount and type of camber change and the amount

of side slip that will occur during suspension jounce and rebound. When the vehicle is in a turn, the outer wheel is put into jounce position and the inner wheel into rebound position due to the transfer of weight caused by centrifugal forces (body roll).

On the design with the instant center outside the vehicle, the outer wheel will have positive camber and the inner wheel will have negative camber. It can be argued that this will reduce tire-to-road contact and therefore reduce good handling in a turn.

On the design with the instant center inside the vehicle, the tendency is to produce negative camber on the outer wheel and positive camber on the inner wheel in a turn. This tends to maintain good tire-to-road contact and therefore good vehicle control.

The instant center on McPherson strut suspension tends to produce negative camber on the outer wheel and positive camber on the inner wheel when the vehicle is in a turn. In this type of suspension, the instant center is determined by the length and angle of the lower control arm and the length and angle of the strut assembly.

The single solid I-beam axle produces

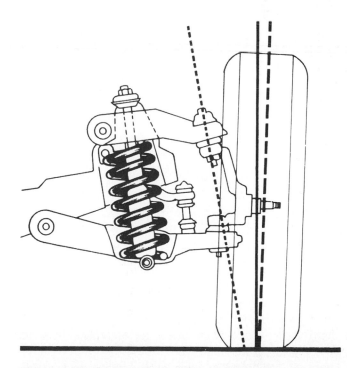

Figure 4-43. Positive scrub radius in long- and short-arm coil-spring front suspension used on rear-wheel-drive vehicle with a front-rear split brake hydraulic system. *(Courtesy of Moog Automotive Inc.)*

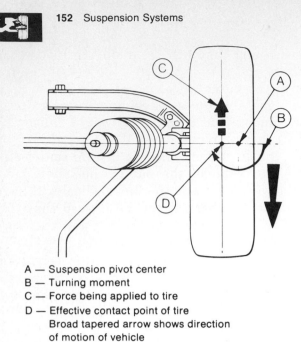

A — Suspension pivot center
B — Turning moment
C — Force being applied to tire
D — Effective contact point of tire
    Broad tapered arrow shows direction
    of motion of vehicle

Figure 4-44. Effect of negative scrub radius during braking on vehicle with diagonally split brake hydraulic system. Force of friction at road surface (C) offsets pull during braking when one half of hydraulic system has failed. *(Courtesy of Ford Motor Co. of Canada Ltd.)*

little camber change and relatively little side slip during jounce and rebound. This is due to the fact that there is a large radius arc produced by the up-and-down movement of the wheel. Twin I-beam axles have a somewhat increased side slip and camber change due to the reduced radius of the arc of wheel travel caused by the pivot point being closer to the wheel.

Side Slip

The amount of side slip or scuff of the tire on the road surface is a result of the camber change as the vehicle suspension moves during jounce and rebound. Side slip is a tire wear factor.

The design objective is a compromise between the effects of more or less side slip. Zero side slip would be best if tire wear were the only consideration. However, a limited amount of side slip helps dampen suspension action due to road surface irregularities.

Suspension design determines the amount of tire scuff or side slip. This is dependent on the relative length and angle of the control arms on long- and short-arm suspension and on the relative length and angle of the control arm and strut on strut-type suspension.

Wheelbase

The wheelbase of a vehicle is the distance between the exact rotating center of the front wheel and the exact center of the rear wheel on the same side of the vehicle, with the wheels in a straight ahead position. The wheelbase of a vehicle has an effect on its ride quality and on its turning radius.

Generally speaking, a longer wheelbase makes for a better riding car---all other factors being equal. This is because the suspension and steering systems have more time to recover from the effects of bumps or holes in the road from the time the front wheels have passed over the bump until the rear wheels pass over the same bump.

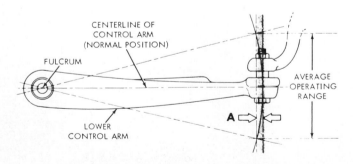

"SIDE-SLIP" — CONTROL ARM IN NORMAL POSITION

A

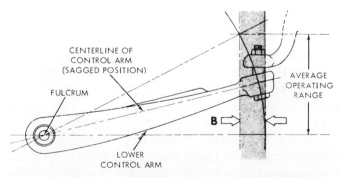

"SIDE-SLIP" — CONTROL ARM IN ABNORMAL POSITION DUE TO SAGGED COIL SPRING

B

Figure 4-45. Normal amount of tire side slip during suspension action with proper suspension height of vehicle shown at A on left. Excessive side slip at B on right as a result of sagged springs. *(Courtesy of Moog Automotive Inc.)*

In general also, the longer the wheelbase, the larger the turning radius required to turn the vehicle around a corner.

The wheelbase of a vehicle may not be exactly the same from one side of the vehicle to the other because of production tolerances and because of differences in front-wheel caster settings. A maximum difference of about 1/4 inch (6.35 millimeters) or as specified by the manufacturer is allowable.

Damage to the suspension system at the front or the rear either through wear or distortion can change the wheelbase of the vehicle. Steering and handling will therefore be adversely affected. The wheelbase of the vehicle is determined by the manufacturer.

Track Width

The track width of a vehicle is the distance measured between the exact center of the tire where it contacts the road and the same point on the tire on the other side of the vehicle, at the front or the rear of the vehicle. The track width of the front wheels and the rear wheels may be the same on some vehicles but may be different at the front as compared to the rear on others. Damage to suspension components will affect track width and vehicle handling.

Understeer and Oversteer

Understeer and *oversteer* are terms used to describe the difference between the direction in which the front wheels are pointed and the actual direction of vehicle movement during a turn.

Understeer is the condition whereby the vehicle is able to turn less than indicated by the turning position of the front wheels. Oversteer is the condition whereby a vehicle makes a sharper turn than indicated by the turning position of the front wheels.

Standard production automobiles are designed to provide some understeer. Since understeer results in slower vehicle response in a turn, it provides a degree of safety for the average driver.

Factors affecting the degree of steering in a vehicle are vehicle design, vehicle speed, body roll, and the ability of the tires to adhere to the road surface. The automotive technician has the most responsibility for this last factor.

Factors affecting the tire's ability to adhere to the road surface and to balanced adhesion between all four tires are correct size and type of tire on all four wheels; condition of tires on all four wheels; proper inflation pressures in all tires; and condition of suspension system including curb height and weight. However, the factor of vehicle speed is the responsibility of the driver, who should take into consideration the condition of the road surface, traffic conditions, and legal speed limits to determine proper vehicle speed.

DIAGNOSTIC CHART: FRONT SUSPENSION

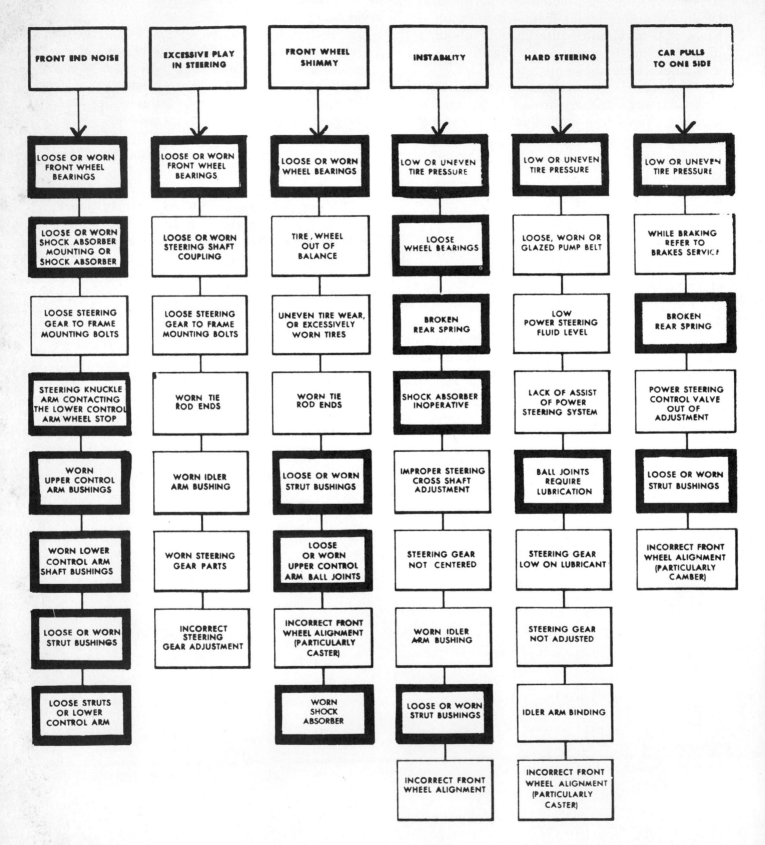

FRONT END NOISE	EXCESSIVE PLAY IN STEERING	FRONT WHEEL SHIMMY	INSTABILITY	HARD STEERING	CAR PULLS TO ONE SIDE
LOOSE OR WORN FRONT WHEEL BEARINGS	LOOSE OR WORN FRONT WHEEL BEARINGS	LOOSE OR WORN WHEEL BEARINGS	LOW OR UNEVEN TIRE PRESSURE	LOW OR UNEVEN TIRE PRESSURE	LOW OR UNEVEN TIRE PRESSURE
LOOSE OR WORN SHOCK ABSORBER MOUNTING OR SHOCK ABSORBER	LOOSE OR WORN STEERING SHAFT COUPLING	TIRE, WHEEL OUT OF BALANCE	LOOSE WHEEL BEARINGS	LOOSE, WORN OR GLAZED PUMP BELT	WHILE BRAKING REFER TO BRAKES SERVICE
LOOSE STEERING GEAR TO FRAME MOUNTING BOLTS	LOOSE STEERING GEAR TO FRAME MOUNTING BOLTS	UNEVEN TIRE WEAR, OR EXCESSIVELY WORN TIRES	BROKEN REAR SPRING	LOW POWER STEERING FLUID LEVEL	BROKEN REAR SPRING
STEERING KNUCKLE ARM CONTACTING THE LOWER CONTROL ARM WHEEL STOP	WORN TIE ROD ENDS	WORN TIE ROD ENDS	SHOCK ABSORBER INOPERATIVE	LACK OF ASSIST OF POWER STEERING SYSTEM	POWER STEERING CONTROL VALVE OUT OF ADJUSTMENT
WORN UPPER CONTROL ARM BUSHINGS	WORN IDLER ARM BUSHING	LOOSE OR WORN STRUT BUSHINGS	IMPROPER STEERING CROSS SHAFT ADJUSTMENT	BALL JOINTS REQUIRE LUBRICATION	LOOSE OR WORN STRUT BUSHINGS
WORN LOWER CONTROL ARM SHAFT BUSHINGS	WORN STEERING GEAR PARTS	LOOSE OR WORN UPPER CONTROL ARM BALL JOINTS	STEERING GEAR NOT CENTERED	STEERING GEAR LOW ON LUBRICANT	INCORRECT FRONT WHEEL ALIGNMENT (PARTICULARLY CAMBER)
LOOSE OR WORN STRUT BUSHINGS	INCORRECT STEERING GEAR ADJUSTMENT	INCORRECT FRONT WHEEL ALIGNMENT (PARTICULARLY CASTER)	WORN IDLER ARM BUSHING	STEERING GEAR NOT ADJUSTED	
LOOSE STRUTS OR LOWER CONTROL ARM		WORN SHOCK ABSORBER	LOOSE OR WORN STRUT BUSHINGS	IDLER ARM BINDING	
			INCORRECT FRONT WHEEL ALIGNMENT	INCORRECT FRONT WHEEL ALIGNMENT (PARTICULARLY CASTER)	

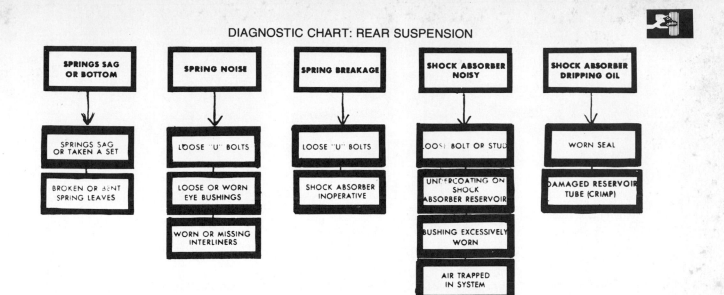

General Precautions

Overhauling suspension systems should not be attempted without being fully aware of all the potential hazards involved. Injury to yourself and others can result if proper procedures and safety measures are not followed. Some of these factors are as follows.

• The vehicle must be properly supported at all times.

• Use only proper tools and supports in good condition and use them as recommended.

• Do not disassemble any suspension parts without proper consideration of the consequences; a loaded spring packs a powerful punch.

• Always refer to the manufacturer's manual for proper procedures; then follow them.

• Never use suspension parts that have been heated, damaged, bent, or straightened.

• Avoid getting grease, solvent, brake fluid, or dirty fingerprints on brake discs, pads, linings, and drums.

• Always tighten all fasteners to specified torque.

• Always install new cotter pins properly locked wherever needed.

Suspension Height

The condition of the vehicle's suspension system, especially the springs, can be determined to some extent by measuring suspension height. This should be done in accordance with manufacturer's procedures and specifications.

In general, the procedure is as follows.

• Vehicle at curb weight (no passengers, no cargo, full tank of fuel, and spare tire in place)

• Vehicle on level floor

• Tire pressures corrected

• Tires of correct size and type

• Accumulations of mud or ice removed

• Measure at points specified by the manufacturer's manual

• Compare measurements to the manufacturer's specifications

• Repair or replace springs in axle pairs only as needed

• Replace suspension parts as needed and as recommended by the manufacturer

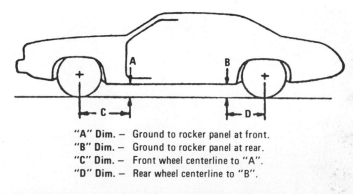

"A" Dim. — Ground to rocker panel at front.
"B" Dim. — Ground to rocker panel at rear.
"C" Dim. — Front wheel centerline to "A".
"D" Dim. — Rear wheel centerline to "B".

Figure 4-46. Typical body suspension height measuring points. Follow shop manual procedures for specific points since these may vary from one make or model to another. *(Courtesy of Moog Automotive Inc.)*

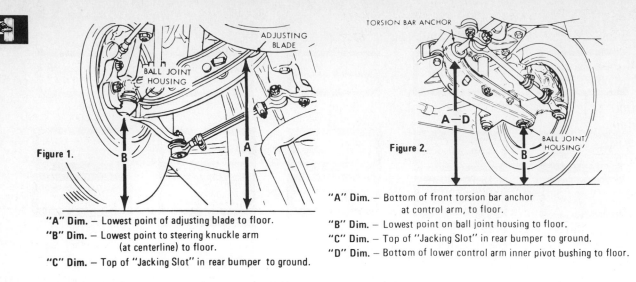

"A" Dim. – Lowest point of adjusting blade to floor.

"B" Dim. – Lowest point to steering knuckle arm
(at centerline) to floor.

"C" Dim. – Top of "Jacking Slot" in rear bumper to ground.

"A" Dim. – Bottom of front torsion bar anchor
at control arm, to floor.

"B" Dim. – Lowest point on ball joint housing to floor.

"C" Dim. – Top of "Jacking Slot" in rear bumper to ground.

"D" Dim. – Bottom of lower control arm inner pivot bushing to floor.

Figure 4-47. Typical height measuring points for torsion bar suspension used on some Chrysler vehicles. Refer to shop manual for specifications and procedures. *(Courtesy of Moog Automotive Inc.)*

Shock Absorbers

A quick check of shock absorber action can be made by jouncing the vehicle as hard as possible at each corner. When the jouncing is stopped, the vehicle should come to rest almost immediately. If not, shock absorbers can be suspected. Visually inspect the shock absorbers for proper mounting, physical damage, or leakage. A slight amount of fluid seepage is normal. Replace damaged or badly leaking shock absorbers in axle pairs. If mounting is faulty or bushings are deteriorated, replace bushings and mount properly. New shock absorbers should be properly bled to remove all air before installation. To prevent preloading of rubber bushings, tighten the shock absorber mounting with vehicle weight on springs.

Air Shock Leak Test

Inflate the shock absorbers to 90 psi (621 kPa) and apply a solution of soapy water to the fittings, lines, and shock absorbers and check for leaks. Leaks will cause the soapy water solution to form bubbles in or near the area of

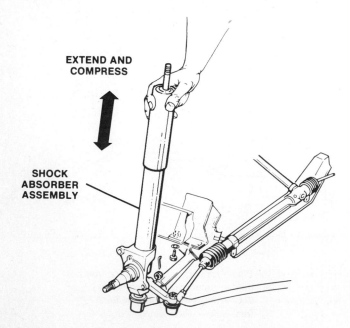

Figure 4-48. Checking shock absorber operation on strut-type suspension. *(Courtesy of Ford Motor Co. of Canada Ltd.)*

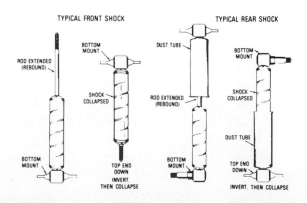

Figure 4-49. Air should be bled from new shock absorbers before installation. With shock in upright position (left above), extend shock fully, then invert shock and collapse completely. Repeat this procedure until all air is bled from the shock. This can be felt when resistance is constant throughout shock piston travel. *(Courtesy of General Motors Corporation)*

leakage. If a leak is detected, repair or replace the defective parts as necessary.

Ball Joints

Ball joint wear should be checked with the vehicle properly supported and ball joints unloaded. Position the dial indicator as recommended by vehicle manufacturer's manual. Grasp the wheel firmly with one hand at the top of the wheel and the other at the bottom. Push in at the top and pull out at the bottom. Alternate this in-and-out movement vigorously and observe total dial indicator needle movement. Compare the reading with the manufacturer's specifications and replace the ball joint if specifications are exceeded. This check is for lateral (radial) movement only.

To check vertical movement, use a bar under the wheel assembly and pry upward; then release. With the dial indicator properly positioned, repeat this movement and observe total dial indicator needle movement. Compare the reading with the manufacturer's specifications and replace the ball joint if specifica-

tions are exceeded. Follow the manufacturer's manual for proper procedure for ball joint replacement. Some ball joints have wear indicators. Follow the manufacturer's directions to determine whether replacement is required.

Control Arms

While the vehicle is supported for checking ball joint wear, check for control arm bushing wear as well. When the wheel is moved laterally, the control arm that is not spring loaded can be checked. The spring-loaded control arm cannot be checked without spring removal except for visual inspection. Damaged or bent control arms must be replaced. Rubber bushings that have cracked must be replaced. Surface checks (very small cracks) are permissible.

Follow the general procedure for spring removal as well as the manufacturer's recommended procedures before removing the control arms. The control arms can then be removed for bushing or arm replacement both front and rear, as specified in shop manual.

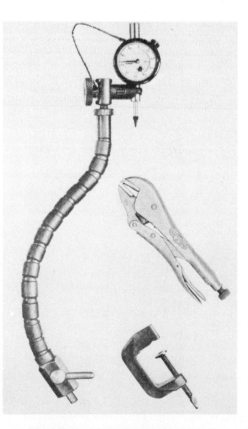

Figure 4-50. Special dial indicator with flexible mounting arm used to check ball joint wear. *(Courtesy of Moog Automotive Inc.)*

Figure 4-51. Axial check of ball joint. For an axial check, first position the dial indicator C-clamp on the control arm, then clean off the flat on the spindle next to the ball joint stud nut. Position the dial indicator on the spindle flat and depress the plunger approximately .250 of an inch. Turn lever to tighten indicator in place. Pry bar between floor and tire and record reading. *(Courtesy of Moog Automotive Inc.)*

Figure 4-52. Radial check of ball joints. For radial check, attach dial indicator to the control arm of the ball joint being checked. Position and adjust plunger of dial indicator against edge of wheel rim nearest to ball joint being checked. Set dial ring to zero marking. Move the wheel in and out and note the amount of ball joint radial looseness registered on the dial. *(Courtesy of Moog Automotive Inc.)*

Strut Rods

If the vehicle is equipped with a strut rod, inspect it for physical damage or distortion. Check bushings for cracks. Replace the rod and bushings as required when servicing control arms.

Springs

Springs should be replaced if suspension height determines springs to be at fault. Leaf springs can be removed and re-arched to restore them to the original position.

Sagging coil springs can be shimmed or coil jacks used to restore vehicle height, but this does nothing to restore the spring. Spring replacement is the only proper solution.

Springs should be replaced according to procedures given in the manufacturer's manual. All component parts required for spring mounting and installation should be inspected and faulty or damaged parts replaced. Tighten all rubber bushing mounting bolts with the vehicle weight on the springs to prevent preloading of bushings. Align the front suspension according to the general procedures in Chapter 7 and according to the manufacturer's manual after any suspension work.

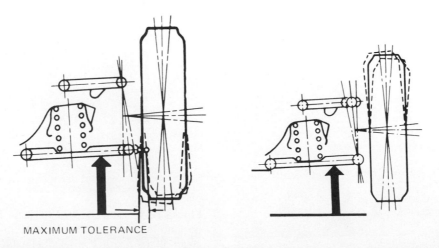

MAXIMUM TOLERANCE

Figure 4-53. Checking ball joint wear on suspension with coil on lower arm. Place jack under lower arm at arrow to unload ball joints fully. Measure vertical and lateral movement in ball joints with dial indicator. If movement exceeds the manufacturer's specifications on any ball joint, it should be replaced. On torsion bar suspension, ball joint wear is measured similarly.

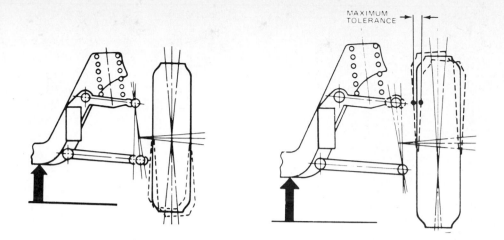

Figure 4-54. Checking ball joint wear on suspension with coil on upper arm. To unload ball joints, block upper arm between arm and frame. Place jack where indicated by arrow and proceed as in Figure 4-53.

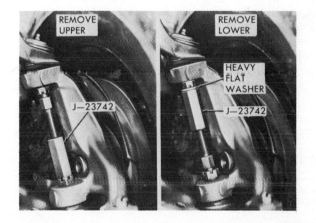

Figure 4-56. Using special tool to force tapered ball joint stud from spindle support. To perform this procedure, ball joint nut must be loosened until flush with end of ball joint stud. Follow the manufacturer's recommended procedures. *(Courtesy of General Motors Corporation)*

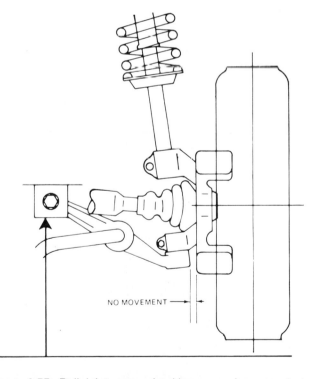

Figure 4-55. Ball joint wear checking procedure on strut suspension. *(Courtesy of Ford Motor Co. of Canada Ltd.)*

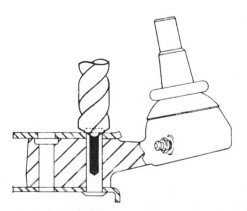

Figure 4-57. On some vehicles, ball joints are riveted to control arm. Rivets should be drilled at exact center to remove head, then drilled with a smaller drill to relieve rivet interference fit in hole. This makes it easier to drive it out of the hole with a pin punch. Replacement ball joint is bolted in.

Figure 4-58. Do not attempt to remove a coil spring until you are aware of the proper procedures and hazards involved. One method of removing a coil spring is shown here. The jack must be handled carefully and the vehicle properly supported to prevent serious injury.

Figure 4-60. Strut-suspension, spring-compressor tool for servicing strut suspension units. *(Courtesy of Moog Automotive Inc.)*

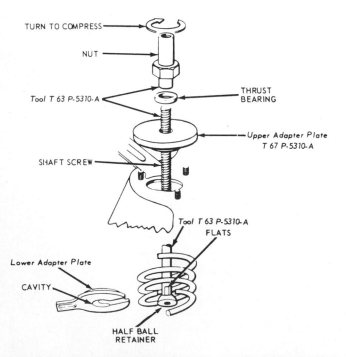

TURN TO COMPRESS

NUT

Tool T 63 P-5310-A

THRUST BEARING

Upper Adapter Plate T 67 P-5310-A

SHAFT SCREW

Tool T 63 P-5310-A

FLATS

Lower Adapter Plate

CAVITY

HALF BALL RETAINER

Figure 4-59. Spring compressor is used to compress spring before removal on suspension with coil on upper arm. Follow the manufacturer's procedures.

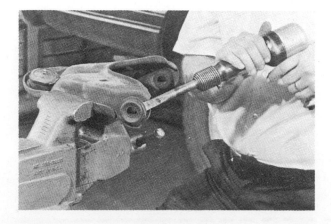

Figure 4-61. Removing rubber control arm bushing with air-operated tool and arm clamped in vise. *(Courtesy of Moog Automotive Inc.)*

USING A NEW BUSHING, SELECT THE DRIVER WHICH BEST FITS THE FLANGE AREA. IT SHOULD CLEAR THE RUBBER PORTION OF THE BUSHING AND NOT EXTEND OVER THE FLANGE AREA OF THE BUSHING, AS SHOWN.

THE NEW BUSHING CAN NOW BE PRESSED IN OR DRIVEN IN PLACE USING A HAMMER.

THIS SET OF DRIVERS WILL ALLOW A PERFECT FIT ON ALMOST ANY AUTOMOTIVE CONTROL ARM BUSHING.

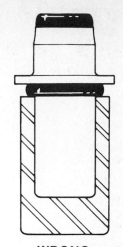

RIGHT
DRIVER IS MAKING FLUSH CONTACT WITH BUSHING RIM.

WRONG
TOO MUCH PLAY BETWEEN DRIVER AND RUBBER SHOULDER.

WRONG
DRIVER DOES NOT MAKE PROPER CONTACT WITH BUSHING RIM.

Figure 4-62. Proper procedure for pressing new suspension bushings into place. *(Courtesy of Moog Automotive Inc.)*

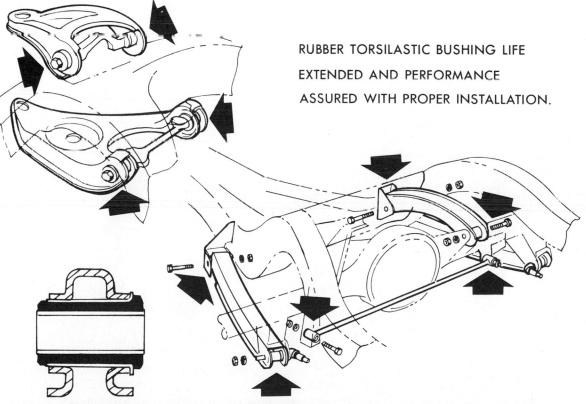

RUBBER TORSILASTIC BUSHING LIFE EXTENDED AND PERFORMANCE ASSURED WITH PROPER INSTALLATION.

Figure 4-63. Rubber torsilastic suspension and steering bushings must be torqued only when the chassis is in its normal "loaded height" position. Rubber torsilastic bushings tightened while the car is on a lift and with wheels hanging free cause high stress, premature failure, and even temporary suspension height changes. Idler arm and center link bushings tightened while wheels are turned may cause the vehicle to pull to one side, hamper returnability, and cause premature failures of the bushings. This applies to all bushings, whether in the front or rear suspensions. Other bushings to consider are the track bar bushings found in various rear coil spring suspensions and front-rear bushings found in leaf springs. Neither should be tightened until the vehicle has been lowered to its normal "loaded height" position. When installing rubber torsilastic bushings, do not press on the rubber or the inner sleeve. Press only on the outer sleeve. *(Courtesy of Moog Automotive Inc.)*

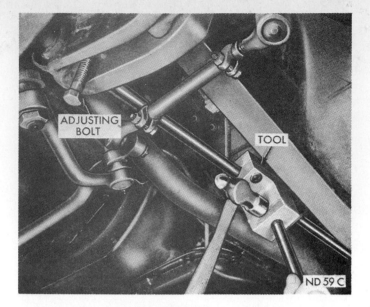

Figure 4-64. To remove torsion bar, support vehicle properly, relieve torsion bar tension at adjusting bolt, remove retaining clips, and drive out torsion bars, using special tool shown. *(Courtesy of Chrysler Corporation)*

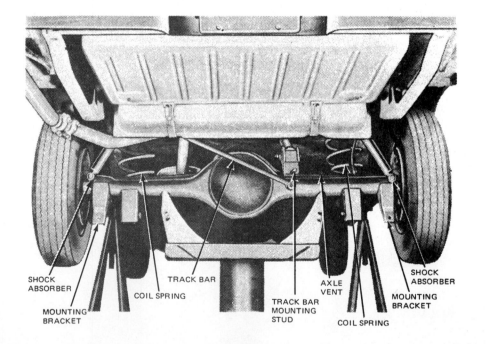

Figure 4-65. This vehicle has been raised on a hoist and the frame supported with jack stands to allow the rear hoist post to raise or lower the rear axle assembly for replacement of springs and other suspension components. *(Courtesy of Ford Motor Co. of Canada Ltd.)*

Stabilizer Bar

The stabilizer bar and bushings can be serviced without spring removal. Replace a damaged or bent bar or linkage. Replace cracked rubber bushings as specified by the shop manual.

Wheel Bearings and Wheels

Inspect and service wheel bearings according to the manufacturer's specifications. Repackable wheel bearings require periodic maintenance, which includes removal, washing in solvent, blowing dry with compressed air, inspection, repacking with wheel-bearing grease, assembly, and adjustment. Factory-sealed bearings require no periodic maintenance.

Refer to Chapter 2 for wheel-bearing service. Always mount wheel assembly by tightening mounting nuts in proper sequence and to specified torque.

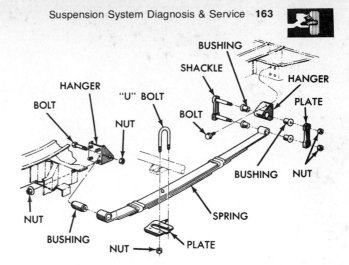

Figure 4-66. Exploded view of leaf-spring mounting with spring hanger and shackle parts shown in detail. *(Courtesy of Chrysler Corporation)*

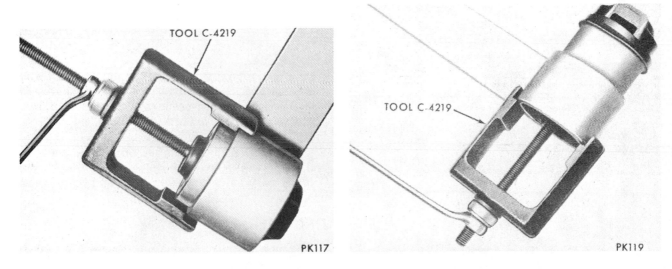

Figure 4-67. A special tool is used to replace some types of leaf-spring eye bushings, as shown here. *(Courtesy of Chrysler Corporation)*

TITLE: <u>FRONT- AND</u>
<u>REAR- SUSPENSION</u>
<u>SYSTEMS</u>

STUDENT'S NAME _____

PERFORMANCE OBJECTIVES

After sufficient opportunity to study this portion of the text and the appropriate training models and with the instructor's supervision and demonstrations, you should be able to perform the following tasks at the request of your instructor.

TASK 1---State the purpose of the suspension system. Describe the construction and operation of the system and its components. Describe the results of normal and abnormal operating conditions on the system's components.

PERFORMANCE EVALUATION

Your instructor may require you to perform these tasks in any of the following ways in order to evaluate your performance:
- By asking test questions
- By asking you to describe the performance of these tasks in writing
- By asking you to describe the performance of these tasks orally

TITLE: <u>SUSPENSION SYSTEM
DIAGNOSIS</u>

STUDENT'S NAME _____

PERFORMANCE OBJECTIVES

After sufficient opportunity to study this portion of the text and the appropriate training models and with the instructor's supervision and demonstrations, you should be able to perform the following tasks at the request of your instructor.

TASK 1---Complete the diagnostic chart provided.

TASK 2---Diagnose suspension system problems on vehicles specified by your instructor.

PERFORMANCE EVALUATION

Your instructor may require you to perform these tasks in any of the following ways in order to evaluate your performance:
- By asking test questions
- By asking you to describe the performance of these tasks in writing
- By asking you to describe the performance of these tasks orally
- By asking you to diagnose suspension system problems on a vehicle specified by your instructor.

PROBLEM	CAUSE	CORRECTION
Front-end noise	1. 2. 3. 4. 5.	1. 2. 3. 4. 5.
Front-wheel shimmy.	1. 2. 3.	1. 2. 3.
Car wanders.	1. 2. 3.	1. 2. 3.
Car pulls to one side.	1. 2. 3.	1. 2. 3.

SUSPENSION SYSTEMS

TITLE: <u>SUSPENSION SYSTEM</u>
<u>SERVICE</u>

STUDENT'S NAME _____

PERFORMANCE OBJECTIVES

After sufficient opportunity to study this portion of the text and the appropriate training models and with the instructor's supervision and demonstrations, you should be able to perform the following tasks at the request of your instructor.

TASK 1--- Overhaul a front- and rear-suspension system on a vehicle specified by your instructor. Follow the procedures and specifications given in the text, in the appropriate shop manual, and by your instructor. Use the worksheet provided to record the results.

PERFORMANCE EVALUATION

Your ability to perform these tasks will be evaluated by your instructor on the basis of how accurately you have followed the procedures and specifications given to you by your instructor, by the text, and by the manufacturer's service manual.

— TRACKING   OK _____   NOT OK _____

— VEHICLE SUSPENSION   OK _____   NOT OK _____
    HEIGHT
    EXPLAIN: LEFT FRONT   _____
              RIGHT FRONT   _____
              LEFT REAR   _____
              RIGHT REAR   _____

— BALL JOINTS   OK _____   NOT OK _____
    EXPLAIN: LEFT UPPER   _____
              LEFT LOWER   _____
              RIGHT UPPER   _____
              RIGHT LOWER   _____

— CONTROL ARMS AND
    BUSHINGS   OK _____   NOT OK _____
    EXPLAIN: LEFT UPPER   _____
              LEFT LOWER   _____
              RIGHT UPPER   _____
              RIGHT LOWER   _____

— STRUT UNITS   OK _____   NOT OK _____
    EXPLAIN: RIGHT   _____
              LEFT   _____

— STABILIZER BAR AND
    LINKS   OK _____   NOT OK _____
    EXPLAIN _____

— SHOCK ABSORBERS   OK _____   NOT OK _____
    EXPLAIN: RIGHT FRONT   _____
              LEFT FRONT   _____
              RIGHT REAR   _____
              LEFT REAR   _____

— REAR SUSPENSION   OK _____   NOT OK _____
    CONTROL ARMS
    EXPLAIN _____

— TRACKING BAR   OK _____ NOT OK _____

EXPLAIN _____

— LEAF SPRINGS
  BUSHINGS   OK _____ NOT OK _____
  SHACKLES   OK _____ NOT OK _____
  REBOUND CLIPS   OK _____ NOT OK _____
  CENTER BOLT   OK _____ NOT OK _____
  U-BOLTS   OK _____ NOT OK _____

— REBOUND RUBBERS
  FRONT   OK _____ NOT OK _____
  REAR   OK _____ NOT OK _____

— COIL SPRINGS
  FRONT   OK _____ NOT OK _____
  REAR   OK _____ NOT OK _____

— ALL APPROPRIATE CORRECTIONS
  COMPLETED ACCORDING TO
  SPECIFICATIONS   YES _____   NO _____

STUDENT SIGN HERE _____

INSTRUCTOR VERIFY _____

Self-Check

1. List three types of automotive frames.
2. List four types of automotive springs.
3. Define Hooke's law.
4. Define jounce and rebound.
5. What is the purpose of the shock absorber?
6. What is the purpose of the stabilizer bar?
7. What is the advantage of independent front suspension over solid axle front suspension?
8. Why is the upper control arm shorter than the lower control arm?
9. Which type of rear suspension uses control arms?
10. What is the purpose of the shackle on leaf springs?
11. Define curb weight.
12. What are the general procedures for checking suspension height? List eight points.
13. How is the amount of ball joint wear determined?
14. Always use _____ cotter pins on all suspension parts.
15. Describe how to bleed air from a hydraulic shock absorber.

Performance Evaluation

After thorough study of this chapter and sufficient practical work on the appropriate components, and with the appropriate shop manual, tools, and equipment, you should be able to do the following.

1. Follow the accepted general precautions.

2. Accurately perform all pre-alignment checks.

3. Accurately measure and correct all alignment factors to the manufacturer's specifications.

4. Properly prepare the wheel assemblies for balancing.

5. Accurately balance the wheel assemblies both statically and dynamically.

6. Perform the necessary checks to determine the success of the alignment and balance procedures.

7. Properly prepare the vehicle for customer acceptance.

8. Complete the Self-Check with at least 80 percent accuracy.

9. Complete all practical work with 100 percent accuracy.

10. Complete all practical work with 100 percent accuracy.

Chapter 5

Wheels and Tires

The tire-and-wheel assemblies provide the only connection between the vehicle and the road. The tires, being air-filled and flexible, absorb much of the road shock from surface irregularities. This reduces the effects of such shock on steering and suspension system components as well as on passengers. In addition, the tires grip the road surface, providing traction for driving, braking, and turning the vehicle.

PART 1   WHEELS

Passenger car wheels consist of a stamped steel disc riveted or welded to a circular rim. Wheel mounting holes are provided in the disc. Mounting holes are tapered to fit tapered mounting nuts that center the wheel over the hub. The rim has a hole for the valve stem and a drop center area for ease of tire removal and installation. The drop center is offset to provide easier tire removal and installation. Other wheels are made of die-cast or forged aluminum or magnesium. Wheels and tires must have minimal radial run-out (out-of-round) and minimal lateral run-out (wobble).

Wheel size is designated as rim width and rim diameter. Rim width is determined by measuring across the rim between the rim flanges. Rim diameter is measured across the bead seating areas from top to bottom of the wheel.

PART 2   TIRES

Most automotive vehicles use the modern tubeless tire. Another type of tire uses an inner rubber tube. This is known as a tube-type tire.

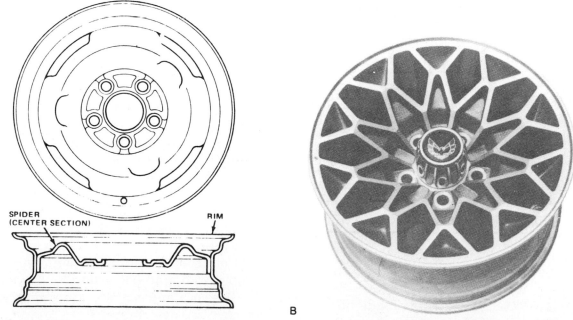

Figure 5-1. Stamped-steel wheel construction detail (A) and cast aluminum wheel (B). (A - *Courtesy American Motors Corporation*; B - *Courtesy of General Motors Corporation*)

A tubeless tire consists of a rubberized cord body attached to two circular beads. An airtight inner layer seals air in the tire. The beads are seated at the outer edges of the wheel rim. A layer of rubber is bonded to the outside of the cord body. This layer of rubber is thickest at the outside circumference of the tire, forming the tire tread.

Tire tread designs include summer- and winter-tread types in many variations. Winter-tread tires are designed to grip the snow and be self-cleaning. Studded tires (steel studs inserted in holes in the tread rubber) provide traction on icy surfaces. Lug spacing is irregular in order to reduce noise as much as possible. Summer tires are designed to provide maximum traction in all conditions. Hydroplaning (tires running on top of water-surfaced road as if water skiing) is reduced by certain tread designs. When tires hydroplane, directional control of the vehicle is lost.

Radial-ply tires provide better traction than bias-ply or bias-belted tires in most circumstances. Radial tires have less rolling resistance and thereby decrease fuel consumption.

Tire sizes are determined by wheel rim diameter and tire width. The correct wheel and tire size combination must be used for any given vehicle. Tire and wheel types and sizes must not be mixed on the same vehicle. Mixing tire and wheel sizes and types affects steering, ride, handling, control, and braking.

Basic Tire Construction*

Every bias-ply, belted-bias, and radial tire shares three basic components: the beads, cord body, and tread.

Tire Beads

The beads in passenger car tires are made of high-tensile steel wires. The beads anchor the tire to the rim. As a tire rotates when traveling on a highway, the force of the tire spinning attempts to throw the tire off the rim. In order to keep the tire from being thrown off the rim at highway speeds, the plies of the tires are attached to rings made of high-tensile steel wire. These rings are designed

---

*Courtesy of Firestone Tire & Rubber Co.

Figure 5-2. Tire cross section showing identification of tire components. *(Courtesy of Firestone Tire and Rubber Co.)*

to fit snugly against the automobile rim and anchor the tire to the rim; these wire rings are referred to as *beads*.

Air pressure in the casing of the tire forces the beads of the tire out against the rim and holds them securely in place. The bead is protected from chafing damage against the rim by a strip of rubber, known as a chafer strip.

Cord Body

The cord body is composed of plies. It retains air under pressure, which supports the load. Layers of fabric or other material laid on top of each other make up the *cord body* of a tire, bonded together in rubber to form one strong unit.

For example, in a conventional four-ply tire the cords in each ply travel on a bias, or diagonally, from the bead on one side of the tire to the bead on the other side. Each layer of plies travels in alternating directions. This bias construction provides the tire with the strength to prevent unnecessary twisting or damage and also allows the cord body to remain flexible enough to absorb shock and provide a smooth, comfortable ride.

The cord angle of a tire is the degree at which the plies cross the center line of any given tire.

Cord body fabrics most commonly used are rayon, nylon, and polyester.

The cord angle in any conventionally constructed tire determines

1. The degree of distortion that may occur at high speeds

2. Handling characteristics

3. Coolness of running

Tire Treads

The tread is applied over the cord body and is then molded into a specific design. The tread comes directly in contact with the road surface.

Separate components are used for the tread area and the sidewall of any tire.

1. *Tread rubber* is compounded, or mixed, to provide customers with a maximum of (a) traction, (b) cut resistance, and (c) long-wearing capabilities.

2. *Sidewall rubber*—In any given tire design, the sidewalls must be flexible and meet different performance requirements than the tread area. Therefore, a different compound is used to produce a suitable rubber for the sidewall area.

Every tire is designed to meet a specific need for performance and safety. All tread designs are developed through an extensive research, development, and testing program, providing quality tires for all types of vehicles and every type of driving—from the stop-and-go driving of the city to the high-speed driving of today's modern highways.

The design of the tread influences:

1. Mileage

2. Handling

3. Ride

4. Amount of road noise

5. Traction

Tread Design

The design of the tread pattern affects the overall performance of any given tire. Depending upon the type of performance required of a tire, the tread will vary in order to meet specific needs. A varying, zigzag tread pattern that provides mileage and traction and reduces side slippage meets most driving requirements.

The tread is molded into a series of grooves and ribs. The ribs provide the traction edges necessary for gripping the road surface and the grooves provide an easy, fast escape for any

foreign matter such as water on the road, and provide the tread edges with a direct, positive grip on the surface being traveled.

A tread rib is the area on which traction edges and sipes are designed. Grooves are the spaces between the adjacent tread ribs.

Sipes

In order to increase the traction ability of any tire and to increase the number of traction edges, small grooves, referred to as sipes, are molded into the ribs of the tread design.

As the tire flexes on the road surface, these sipes open and provide extra gripping action.

When studying a tread surface, the sipes look as though they are very shallow and only on the surface. The sipes travel the full depth of the tread providing traction for the entire tread life of the tire.

The more grooves and sipes a tire has, the greater its wet pavement traction.

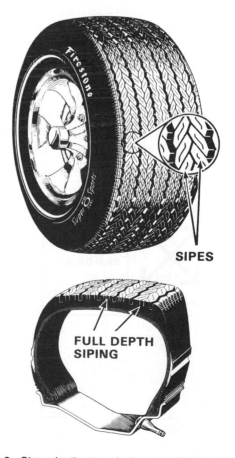

SIPES

FULL DEPTH SIPING

Figure 5-3. Sipes in tire treads improve the tractive ability of a tire on wet surfaces. *(Courtesy of Firestone Tire and Rubber Co.)*

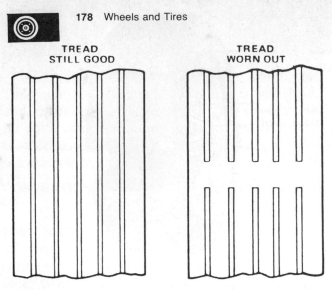

TREAD STILL GOOD   TREAD WORN OUT

Figure 5-6. Tread pattern on tires has several sections where grooves are not as deep as elsewhere. These sections are called tread wear bars. When tire tread is worn down to tread wear bars, tire should be replaced. *(Courtesy of American Motors Corporation.)*

In *radial* tires the belt plies eliminate any contraction or expansion of the tread design, as illustrated in Figure 5-8 at the right. The absence of the scrubbing action, caused by contraction and expansion, makes radial tires the ultimate for long mileage. The unique flex action of the radial tire as it goes through the footprint cycle gives a longer footprint, as shown in the contraction cycle.*

Tread Wear Indicators

As a visible check of tire condition, tread wear indicators are molded into the bottom of the tread grooves. These indicators appear in the form of 1/2-inch (13 millimeters) wide bands across the tread when it is worn to a thickness of 1/16-inch (1.58 millimeters). The tire should be replaced when these indicator bands become visible. A number of states have statutes concerning minimum permissible tread depths and use these indicators as the tire wear limit.

Slip Angle

The slip angle of a tire is the difference between the direction in which a wheel is positioned and the direction in which it is actually moving. As the lateral load on a tire is increased during a turn, the slip angle increases. If the lateral load is increased sufficiently, the tire will actually slip sidewise.

---

*Courtesy of Firestone Tire & Rubber Co.

Bias-ply tires have the greater slip angle. Bias-belted tires have a lower slip angle, and radial-ply tires have the lowest slip angle. As the slip angle increases, control of the vehicle becomes more difficult. Adhesion of the tire to the road surface (coefficient of friction) is critical to vehicle control.

The rate of slip-angle increase (until the point of loss of control) is more gradual with bias-ply tires as compared to radial-ply tires. As higher speed and sharper turns increase the lateral force on the tires, the radial-ply tire suddenly loses adhesion with the road surface and side slip results. A bias-ply tire, on the other hand, has a gradual increase in slip angle, which provides more warning time to the driver of an impending slip.

Factors affecting a tire's slip angle are lateral force on the tire, depending on vehicle speed; load and degree of turn; tire inflation pressure; tire temperature; type of tire rubber compound; type of tire construction (bias-ply, bias-belted, radial); tire condition (wear); type of the road surface (pavement, gravel, etc.); and condition of the road surface (wet, dry, icy).

It is extremely important that (a) a vehicle be equipped with the recommended tire type and size, (b) only tires in good condition be used, (c) proper inflation pressures be maintained, and (d) tire types and sizes not be mixed in order to maintain good vehicle handling and safety of directional control and braking.

Radial-Tire Lead and Waddle

Lead is the tendency of the car to deviate from a straight path on a level road when there

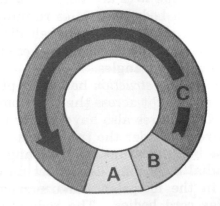

Figure 5-7. A car tire goes through three stress cycles during every revolution of the wheel. A - contraction, B - expansion, C - normal stress. *(Courtesy of Firestone Tire and Rubber Company)*

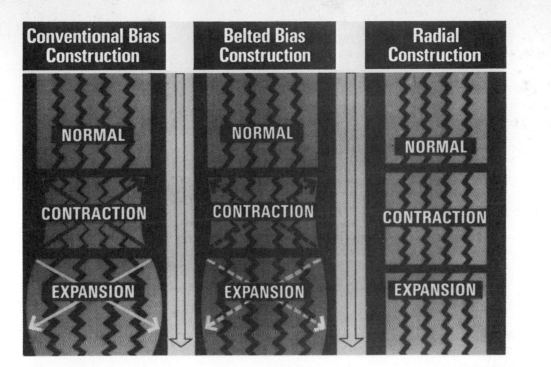

Figure 5-8. Effects of stress cycles on tire tread of tires of different construction. Radial tires show least effect. *(Courtesy of Firestone Tire and Rubber Company)*

is no pressure on the steering wheel in either direction. Lead may be caused by improper alignment, uneven braking, or the tires.

An off-center belt in a radial tire can cause a tire to lead. This type of tire creates a side force on the vehicle resulting in lead. Also, if one side of a tire is smaller in diameter than the other side, the tire will cause the vehicle to lead. Front tires with unequal tread wear can also cause lead. Tires must be of the same size, type, and condition on the front wheels to prevent tire lead. Rear tires do not cause lead; however, rear tires should also be of the same size, type, and condition to assure balanced braking and vehicle control.

Load Rating

The load-carrying ability of a tire is indicated by a letter designation on the sidewall of the tire. Letter ratings of A, B, C, D, etc., indicate progressively higher load ratings. The load-rating D tire is capable of carrying a greater load than a C load rating tire. Refer to the load-rating chart (Figure 5-16) for comparisons. Load ratings are given for specific inflation pressures only. However, maximum inflation pressures recommended by the tire manufacturer should not be exceeded. Tire failure resulting in accidents and injury can be the result.

Tire Inflation Pressures

Tire inflation pressures are stated in pounds per square inch (psi) or in kilopascals (kPa). Maximum inflation pressures are usually stated on the sidewall of the tire and are to be used when maximum stated load on the same tire is to be carried. Lesser pressures may be used as recommended by the vehicle manufacturer with lighter loads. The pressure recommended for any model is carefully calculated to give a satisfactory ride, stability, steering, tread wear, tire life, and resistance to bruises.

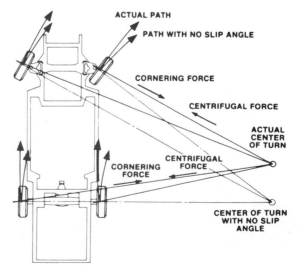

Figure 5-9. Effects of tire slip angle. *(Courtesy of Ford Motor Co.)*

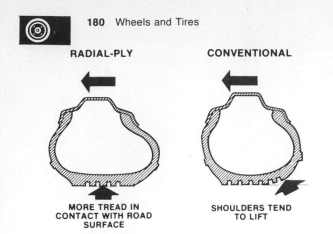

RADIAL-PLY   **CONVENTIONAL**

MORE TREAD IN CONTACT WITH ROAD SURFACE

SHOULDERS TEND TO LIFT

Figure 5-10. More flexible sidewalls and radial-ply construction improve tire adhesion to road surface as compared to conventional tire construction during a turn. This reduces the amount of slip angle. *(Courtesy of Ford Motor Co.)*

Tire pressure, with tires cold (after the car has sat for three hours or more or driven less than one mile), should be checked monthly or before any extended trip and set to the specifications on the tire placard located on rear face of driver's door, door post, or glove compartment. Valve caps or extensions should be on the valves to keep dust and water out.

Tire pressures may increase as much as 40 kPa (6 psi) when hot. Higher-than-recommended pressure can cause

1. Hard ride
2. Tire bruising or carcass damage
3. Rapid tread wear at center of tire

Lower-than-recommended pressure can cause

1. Tire squeal on turns
2. Hard steering
3. Rapid and uneven wear on the edges of the tread
4. Tire rim bruises and rupture
5. Tire cord breakage
6. High tire temperatures

Figure 5-11. Effects of radial tire waddle on vehicle. *(Courtesy of General Motors Corporation)*

7. Reduced handling
8. High fuel consumption

Unequal pressure on the same axle can cause

1. Uneven braking
2. Steering lead
3. Reduced handling
4. Swerve on acceleration

Tire Sizes

Tire size designations are stated on the tire sidewall. Tire cross-sectional width and bead-to-bead diameter are the major size designations. Typical sidewall tire size and type markings are shown in Figures 5-13 and 5-15.

Radial Tire Grading

Radial tires produced by the various tire manufacturers are classified according to their tread life, wet traction capability, and high-temperature capability. The high-temperature capability is directly related to the tire's high-speed capability. These ratings are moulded into the tire sidewall of some tires.

The tread wear rating is stated in numbers of increments of ten. Every ten points represents 3,000 miles (4800 km) of tread life expectancy. A rating of 100 for tread wear would therefore indicate a tread life expectancy of 30,000 miles (48,000 km). Traction ratings and temperature resistance ratings are stated in letter grades A, B and C, with A the best.

INFLATION PRESSURE CONVERSION CHART (KILOPASCALS TO PSI)			
kPa	psi	kPa	psi
140	20	215	31
145	21	220	32
155	22	230	33
160	23	235	34
165	24	240	35
170	25	250	36
180	26	275	40
185	27	310	45
190	28	345	50
200	29	380	55
205	30	415	60
Conversion: 6.9 kPa = 1 psi			

Figure 5-12. Tire pressure conversion chart. *(Courtesy of General Motors Corportion)*

The compact spare is designed to reduce weight and increase trunk space. It will perform well with the other tires on the car for the 1,600 to 4,800 km (1,000 to 3,000 miles) life of the tread. A narrow wheel is used. The compact spare tire should not be used with any other wheel. The compact spare wheel should not be used with the standard tires, snow tires, wheel covers, or trim rings. If such use is attempted, damage to these items or other parts of the car may occur. The compact spare should only be used on cars that offered it as original equipment. Since it is smaller in diameter, it must not be used on cars equipped with a limited slip differential, because possible damage to the differential might occur. Inflation pressure should be periodically checked and maintained at recommended pressure. The compact spare is serviceable with present equipment and procedures.

The compact spare should be used only as recommended by the vehicle manufacturer.

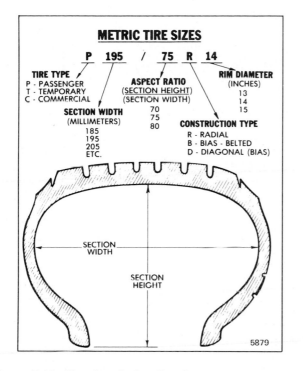

Figure 5-13. Tire size designation found on tire sidewall is explained here. *(Courtesy of General Motors Corporation)*

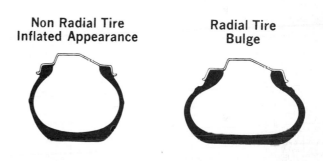

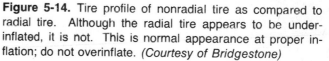

Figure 5-14. Tire profile of nonradial tire as compared to radial tire. Although the radial tire appears to be underinflated, it is not. This is normal appearance at proper inflation; do not overinflate. *(Courtesy of Bridgestone)*

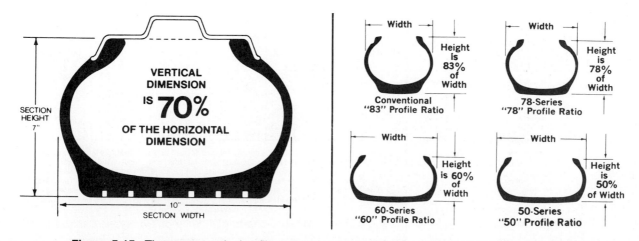

Figure 5-15. The aspect ratio (profile ratio) of a tire is its cross-sectional height compared to its cross-sectional width expressed in precentage figures.

TIRE SIZE OR DESIGNATION

LOAD LIMITS (LBS. PER TIRE) FOR TIRES USED ON PASSENGER CARS, STATION WAGONS AND MULTI-PURPOSE PASSENGER VEHICLES

Load Range B (4 ply-rating) ⟶
Load Range C (6-ply rating) ⟶
Load Range D (8-ply rating) ⟶

Conventional Bias Ply 1965-On	78 Series	70 Series	60 Series	50 Series	Metric	Radial Ply 78 Series	70 Series	20	22	24	26	28	30	32	34	36	38	40
6.00-13					165R13			770	820	860	900	930	970	1010	1040	1080	1110	1140
	A78-13	A70-13	A60-13			AR78-13	AR70-13	810	860	900	940	980	1020	1060	1090	1130	1160	1200
6.50-13	B78-13		B60-13	B50-13	175R13	BR78-13	BR70-13	890	930	980	1030	1070	1110	1150	1190	1230	1270	1300
	C78-13	C70-13	C60-13			CR78-13	CR70-13	950	1000	1050	1100	1140	1190	1230	1270	1320	1360	1400
7.00-13								980	1030	1080	1130	1180	1230	1270	1310	1360	1400	1440
					185R13			950	1000	1050	1100	1140	1190	1230	1270	1320	1350	1400
	D78-13	D70-13	D60-13					1010	1070	1120	1170	1220	1270	1320	1360	1410	1450	1490
					195R13			1060	1110	1170	1220	1280	1320	1370	1420	1470	1510	1560
6.45-14					155R14			780	820	860	900	940	970	1010	1040	1080	1110	1140
	B78-14							860	910	960	1000	1040	1080	1120	1160	1200	1240	1270
						BR78-14		870	930	980	1030	1070	1110	1150	1190	1230	1270	1300
					165R14			860	910	960	1000	1040	1080	1120	1170	1200	1240	1280
6.95-14								950	1000	1050	1100	1140	1190	1230	1270	1310	1350	1390
	C78-14					CR78-14	CR70-14	950	1000	1050	1100	1140	1190	1230	1270	1320	1360	1400
					175R14			950	1000	1050	1100	1140	1190	1230	1280	1320	1360	1400
	D78-14	D70-14	D60-14			DR78-14	DR70-14	1010	1070	1120	1170	1220	1270	1320	1360	1410	1450	1490
7.35-14								1040	1100	1160	1210	1260	1310	1360	1400	1450	1490	1540
					185R14			1040	1100	1160	1210	1260	1310	1360	1410	1450	1500	1540
	E78-14	E70-14	E60-14			ER78-14	ER70-14	1070	1130	1190	1240	1300	1350	1400	1440	1490	1540	1580
7.75-14								1150	1210	1270	1330	1390	1440	1500	1550	1600	1650	1690
					195R14			1150	1210	1270	1330	1390	1440	1500	1540	1590	1640	1690
	F78-14	F70-14	F60-14			FR78-14	FR70-14	1160	1220	1280	1340	1400	1450	1500	1550	1610	1650	1700
8.25-14								1250	1310	1380	1440	1500	1560	1620	1670	1730	1780	1830
					205R14			1250	1310	1380	1440	1500	1560	1620	1680	1730	1780	1830
	G78-14	G70-14	G60-14	G50-14		GR78-14	GR70-14	1250	1310	1380	1440	1500	1560	1620	1680	1730	1780	1830
8.55-14								1360	1430	1510	1580	1640	1710	1770	1830	1890	1950	2000
					215R14			1360	1430	1510	1580	1640	1710	1770	1830	1890	1950	2010
	H78-14	H70-14	H60-14	H50-14		HR78-14	HR70-14	1360	1440	1510	1580	1650	1710	1770	1830	1890	1950	2010
8.85-14								1430	1510	1580	1660	1730	1790	1860	1920	1990	2050	2100
	J78-14	J70-14	J60-14			JR78-14	JR70-14	1430	1500	1580	1650	1720	1790	1860	1920	1980	2040	2100
					225R14			1430	1510	1580	1660	1730	1790	1860	1920	1980	2040	2100
		L70-14	L60-14				LR70-14	1520	1600	1680	1750	1830	1900	1970	2040	2100	2170	2230
				M50-14				1610	1700	1780	1860	1940	2020	2090	2160	2230	2300	2370
				N50-14				1700	1790	1880	1970	2050	2130	2210	2280	2360	2430	2500
	A78-15	A70-15				AR78-15		810	860	900	940	980	1020	1060	1090	1130	1160	1200
					165R15	BR78-15		870	910	960	1000	1050	1090	1130	1170	1200	1240	1280
			B60-15					890	930	980	1030	1070	1110	1150	1190	1230	1270	1300
6.85-15	C78-15	C70-15						950	1000	1050	1100	1140	1190	1230	1270	1320	1360	1390
			C60-15					950	1000	1050	1100	1140	1190	1230	1270	1320	1360	1400
					175R15			950	1000	1050	1100	1140	1190	1230	1280	1320	1360	1400
	D78-15	D70-15					DR70-15	1010	1070	1120	1170	1220	1270	1320	1360	1410	1450	1490
7.35-15								1070	1130	1180	1240	1290	1340	1390	1440	1480	1530	1570
	E78-15	E70-15	E60-15			ER78-15	ER70-15	1070	1130	1190	1240	1300	1350	1400	1440	1490	1540	1580
					185R15			1070	1130	1180	1240	1290	1340	1390	1430	1480	1520	1570
7.75-15					195R15			1150	1210	1270	1330	1380	1440	1490	1540	1590	1640	1690
	F78-15	F70-15	F60-15			FR78-15	FR70-15	1160	1220	1280	1340	1400	1450	1500	1550	1610	1650	1700
8.15-15					205R15			1240	1300	1370	1430	1490	1550	1610	1660	1720	1770	1820
	G78-15	G70-15	G60-15			GR78-15	GR70-15	1250	1310	1380	1440	1500	1560	1620	1680	1730	1780	1830
8.25-15								1250	1310	1380	1440	1500	1560	1620	1670	1730	1780	1830
8.45-15								1340	1410	1480	1550	1620	1680	1740	1800	1860	1920	1970
					215R15			1340	1410	1480	1550	1620	1680	1740	1800	1860	1910	1970
	H78-15	H70-15	H60-15			HR78-15	HR70-15	1310	1400	1450	1520	1580	1640	1710	1760	1820	1880	1930
8.55-15								1360	1440	1510	1580	1650	1710	1770	1830	1890	1950	2010
								1360	1430	1510	1580	1640	1710	1770	1830	1890	1950	2000
8.85-15					225R15			1430	1510	1580	1650	1720	1790	1860	1920	1980	2040	2100
	J78-15	J70-15	J60-15			JR78-15	JR70-15	1430	1500	1580	1650	1720	1790	1860	1920	1980	2040	2100
9.00-15								1460	1540	1620	1690	1760	1830	1900	1970	2030	2090	2150
		K70-15					KR70-15	1460	1540	1620	1690	1770	1830	1900	1970	2030	2090	2150
9.15-15								1510	1600	1680	1750	1830	1900	1970	2030	2100	2160	2230
					235R15			1510	1600	1680	1750	1830	1900	1970	2040	2100	2170	2230
	L78-15	L70-15	L60-15			LR78-15	LR70-15	1520	1600	1680	1750	1830	1900	1970	2040	2100	2170	2230
						MR78-15	MR70-15	1610	1700	1780	1860	1940	2020	2090	2160	2230	2300	2370
	N78-15					NR78-15		1700	1790	1880	1970	2050	2130	2210	2280	2360	2430	2500
8.90-15								1700	1810	1880	1970	2050	2130	2210	2290	2360	2430	2500
6.00-16								1075	1135	1195	1250	1300	1350	1400	1450	1500		
6.50-16								1215	1280	1345	1405	1465	1525	1580	1635	1690	1740	1790
7.00-15								1310	1380	1450	1515	1580	1640	1700	1760	1820	1870	1930
7.00-16								1365	1440	1515	1585	1650	1715	1780	1840	1900		

Maximum inflation and load is that shown in the 32 psi column for Load Range B (4-ply rating) tires, 36 psi for Load Range C (6-ply rating) tires and 40 psi for Load Range D (8-ply rating) tires.

If you are considering replacing present tires with tires of a different size designation, be sure to check the automobile manufacturer's recommendations. Interchangeability is not always possible because of differences in load ratings, tire dimensions, wheel well clearances and rim sizes. Also, tires of different construction (bias, bias/belted, or radial) or different sizes should never be used together on the same axle. If radial tires are used with other tire types, the radials must be used on the rear axle only.

Figure 5-16. Tire size, load rating, and inflation pressure chart. *(Courtesy of Goodyear Canada Inc.)*

DIAGNOSTIC CHART

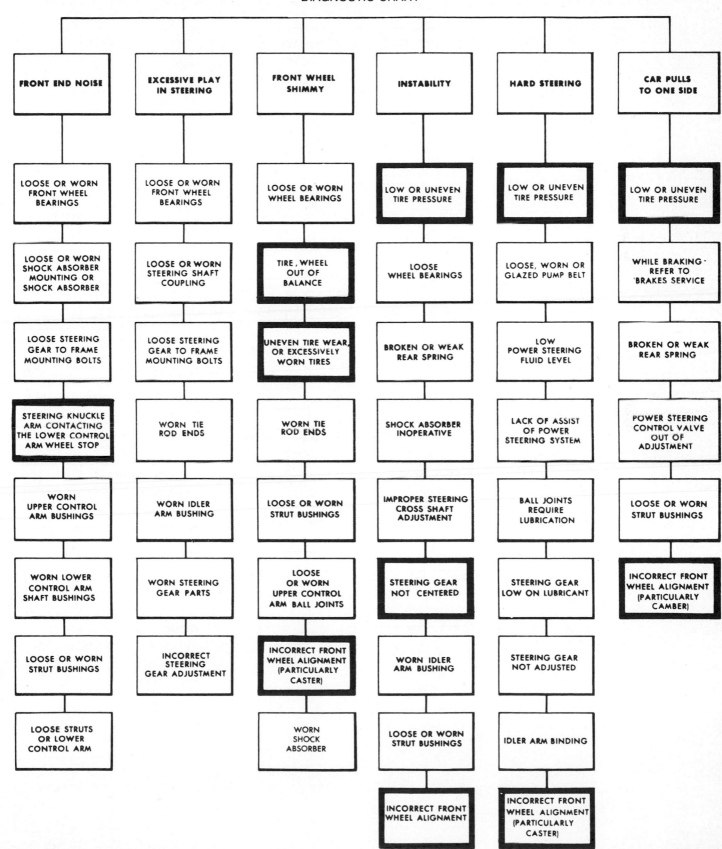

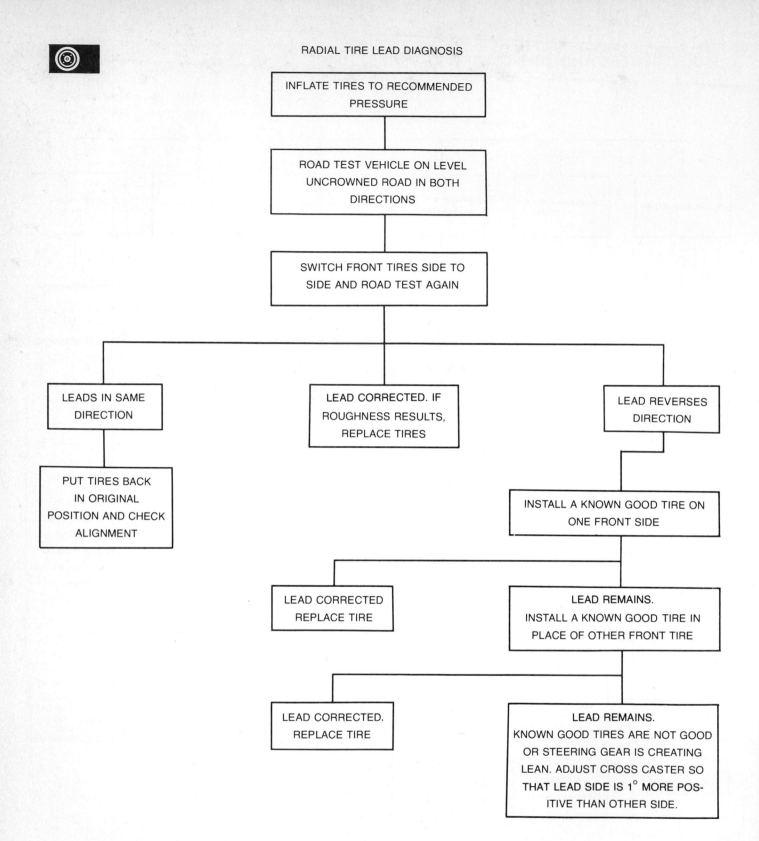

INFLATE TIRES TO RECOMMENDED PRESSURE

ROAD TEST VEHICLE ON LEVEL UNCROWNED ROAD IN BOTH DIRECTIONS

SWITCH FRONT TIRES SIDE TO SIDE AND ROAD TEST AGAIN

LEADS IN SAME DIRECTION

LEAD CORRECTED. IF ROUGHNESS RESULTS, REPLACE TIRES

LEAD REVERSES DIRECTION

PUT TIRES BACK IN ORIGINAL POSITION AND CHECK ALIGNMENT

INSTALL A KNOWN GOOD TIRE ON ONE FRONT SIDE

LEAD CORRECTED REPLACE TIRE

LEAD REMAINS. INSTALL A KNOWN GOOD TIRE IN PLACE OF OTHER FRONT TIRE

LEAD CORRECTED. REPLACE TIRE

LEAD REMAINS. KNOWN GOOD TIRES ARE NOT GOOD OR STEERING GEAR IS CREATING LEAN. ADJUST CROSS CASTER SO THAT LEAD SIDE IS 1° MORE POS- ITIVE THAN OTHER SIDE.

General

Vibration, roughness, tramp, shimmy and thump may be caused by excessive tire or wheel run-out, worn or cupped tires, or wheel and tire unbalance. These problem conditions may also be caused by rough or undulating road surfaces. Driving the automobile on different types of road surfaces will indicate if the road surfaces are actually causing the problem.

Always road-test the automobile, preferably with the owner in the automobile, to determine the exact nature of the problem. The automobile should be driven at least 7 miles to warm the tires and remove flat spots that may have formed temporarily while the automobile was parked. Note tire condition and wear, and check and adjust tire inflation pressures before road testing.

Radial-Tire Performance Characteristics

Because of their unique construction, radial-ply tires produce ride, handling, and appearance characteristics noticeably different from conventional tires. Radial-ply tire ride quality and feel may seem harsh, particularly at low speeds. This is due to the stiff belts used in the construction of these tires. Harshness often leads to the assumption that the tires are overinflated. Inflate radial-ply tires to recommended levels only.

_____

*Courtesy of American Motors.

Radial-ply tires have a highly flexible sidewall, which produces a characteristic sidewall bulge, making the tire appear under-inflated. This is a normal condition for radial-ply tires. Do not attempt to reduce this bulge by overinflating the tire. Always check tire inflation pressures, using an accurate gauge, and inflate the tires to recommended levels only.

Radial-ply tires also produce a side-to-side or waddle motion that is most noticeable at speeds of 15 miles per hour (mph) (24 km/h) or less. This motion is a normal characteristic of radial-ply tires and is a result of their unique construction. An objectionable waddle condition can sometimes be reduced by rotating the tires front-to-rear; however, do not attempt to correct a waddle condition by balancing.

Proper mounting and balancing of radial-ply tires is very important. Improper balancing or incomplete seating of the tire bead can produce a high frequency vibration noticeable throughout the automobile at speeds above approximately 45 mph (72 km/h). Improper bead seating can be checked by visually inspecting the tire. To correct unbalance, reseat the bead if necessary, and balance the tire using dynamic, two-plane balancing equipment. This type of balancing equipment is essential to solving radial-ply tire unbalance problems.

Tire Thump

Thump is a noise caused by the tire moving over irregularities in the road or by irregularities within the tire itself. The thump

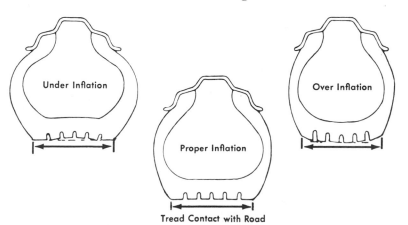

Figure 5-17. Effects of different inflation pressures on the tire's footprint. Handling, control, ride quality, and tire life are all affected by inflation pressures. Always inflate tires to recommended pressures only.

sound will coincide with each wheel revolution.

To determine which tire is causing thump, temporarily inflate all tires to 50 psi (345 kPa) and drive over the same roads. If this procedure eliminates the problem, reduce the air pressure in one tire at a time and repeat the road test. Perform this procedure until all tires have been tested and each test is made with three tires at high pressure and one tire at recommended pressure. When thump again develops, the tire just deflated to the recommended pressure is the defective tire and should be replaced. *Note*: Although the procedure for diagnosing tire thump is quite effective with conventional tires, it is considerably less effective with radial tires.

Tire Tramp

Tire tramp is caused by tire and wheel static unbalance or by excessive radial and lateral run-out of the tire or wheel. The most effective method for checking tire and wheel static balance is by using off-the-automobile balancing equipment.

Static balance is the result of an equal distribution of wheel and tire weight about the spindle in such a manner that the assembly lacks the tendency to rotate by itself when mounted on the arbor of a balancing machine. Static unbalance occurs when an unequal portion of weight is concentrated at one point on the tire and wheel. It causes a vibratory-type pounding action, which is referred to as tire tramp, wheel tramp, or wheel hop.

Dynamic balance is the result of an equal distribution of wheel and tire weight around the plane of rotation, which causes the wheel to rotate smoothly about the axis that bisects the wheel and tire centerline. Dynamic unbalance occurs when unequal forces are concentrated at opposing points on the tire circumference. It causes wheel shimmy and vibration at medium and high speeds.

The most effective methods for balancing wheels and tires is by using equipment that will correct both static and dynamic balance conditions. Dynamic, two-plane balancing equipment is preferable. Since procedures vary with different machines, follow the equipment manufacturer's instructions explicitly.

Figure 5-18. Several tire wear indicator bars are located around the tire tread. When tread wear exposes these fully, there is only about 1.6 millimeters of tread remaining. Tire should then be replaced. *(Courtesy of Bridgestone)*

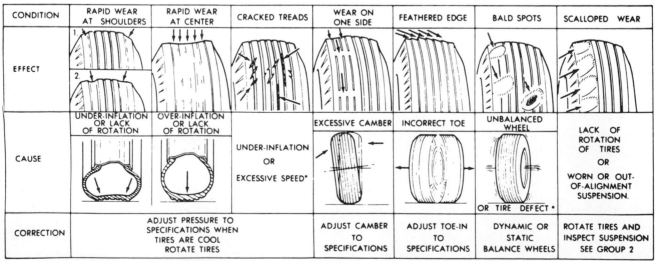

CONDITION	RAPID WEAR AT SHOULDERS	RAPID WEAR AT CENTER	CRACKED TREADS	WEAR ON ONE SIDE	FEATHERED EDGE	BALD SPOTS	SCALLOPED WEAR
EFFECT							
CAUSE	UNDER-INFLATION OR LACK OF ROTATION	OVER-INFLATION OR LACK OF ROTATION	UNDER-INFLATION OR EXCESSIVE SPEED*	EXCESSIVE CAMBER	INCORRECT TOE	UNBALANCED WHEEL OR TIRE DEFECT*	LACK OF ROTATION OF TIRES OR WORN OR OUT-OF-ALIGNMENT SUSPENSION.
CORRECTION	ADJUST PRESSURE TO SPECIFICATIONS WHEN TIRES ARE COOL ROTATE TIRES			ADJUST CAMBER TO SPECIFICATIONS	ADJUST TOE-IN TO SPECIFICATIONS	DYNAMIC OR STATIC BALANCE WHEELS	ROTATE TIRES AND INSPECT SUSPENSION SEE GROUP 2

*HAVE TIRE INSPECTED FOR FURTHER USE.

Figure 5-19. Tire wear conditions are stated across the top. The resulting effects are illustrated in the next column across. Causes of these conditions are shown next, with the appropriate corrective procedures across the bottom. *(Courtesy of Chrysler Corporation)*

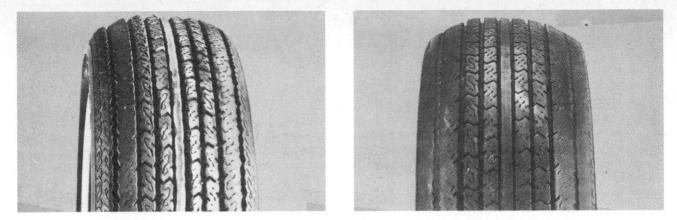

Figure 5-20. Example of tread wear (feather-edge tread) due to excessive toe-in or toe-out at left. Two-sided wear can be the result of underinflation or high-speed cornering.

Figure 5-21. One-side wear at left indicates excessive camber. Cupped wear at right is the result of underinflation or imbalance.

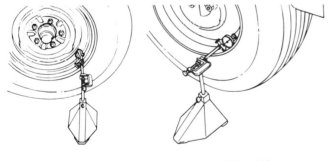

RADIALLY   LATERALLY

Figure 5-22. Using a dial indicator to measure wheel run-out. *(Courtesy of American Motors Corporation)*

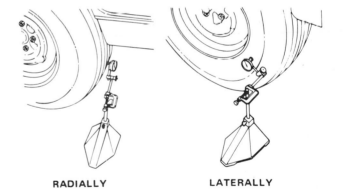

RADIALLY   LATERALLY

Figure 5-23. Using a dial indicator to measure tire run-out. *(Courtesy of American Motors Corporation)*

Wheel and Tire Run-out

Excessive radial and lateral run-out of a wheel-and-tire assembly can cause roughness, vibration, tramp, tire wear, and steering wheel tremor.

Before checking run-out and to avoid false readings caused by temporary flat spots in the tires, check run-out only after the automobile has been driven at least 7 miles (11 km).

The extent of run-out should be measured with a dial indicator. All measurements should be made on the automobile with the tires inflated to recommended reduced load inflation pressures and with the wheel bearings adjusted to specifications.

Measure tire radial run-out at the center and outside ribs of the tread face. Measure tire lateral run-out just above the buffing rib on the sidewall. Mark the high points of lateral and radial run-out for future reference. On conventional tires, radial run-out must not exceed 0.105 inch (2.66 millimeters) and lateral run-out must not exceed 0.100 inch (2.54 millimeters). On radial-ply tires, radial run-out must not exceed 0.080 inch (2.03 millimeters) and lateral run-out must not exceed 0.100 inch (2.54 millimeters).

If total radial or lateral run-out of the tire exceeds specified limits, it will then be necessary to check wheel run-out to determine whether the wheel or tire is at fault. Wheel radial run-out is measured at the wheel rim just inside the wheel cover retaining nibs. Wheel lateral run-out is measured at the wheel rim bead flange just inside the curved lip of the flange. Wheel radial run-out should not exceed 0.035 inch (0.89 millimeter) and wheel lateral run-out should not exceed 0.045 inch (1.14 millimeters). Mark the high points of radial and lateral run-out for future reference.

If total tire run-out, either lateral or radial, exceeds the specified limit but wheel run-out is within the specified limit, it may be possible to reduce run-out to an acceptable level by changing the position of the tire on the wheel so that the previously marked high points are 180° apart.

Vibration

General

Vibration may be caused by tire/wheel unbalance or run-out, incorrectly adjusted wheel bearings, loose or worn suspension or steering components, worn or defective tires, incorrect universal joint angles, worn universal joints, excessive propeller shaft or yoke run-out, rotor or brakedrum run-out, loose engine or transmission supports, or engine-driven accessories.

Vibration Categories

Vibrations can be divided into two categories: mechanical and audible. Mechanical vibrations are felt through the seats, floorpan, or steering wheel and usually produce some visible motion in the rear-view mirror, front fenders, dash panel, or steering wheel.

Audible vibrations are heard or sometimes sensed above normal road and background noise and may be accompanied by a mechanical vibration. In some cases, they occur as a droning or drumming noise. In other cases, they produce a buffeting sensation felt or sensed by the driver rather than heard.

Vibration Sensitivity

Mechanical and audible vibrations are sensitive to changes in engine torque, automobile speed, or engine speed. They usually occur within one and sometimes two well-defined ranges in terms of automobile speed, engine revolutions per minute (rpm), and torque application.

Consider for correction only those items coded on the charts that are related to the problem condition. Refer to the correction codes for a definition of the various correction procedures.

Vibration Diagnosis Chart Codes

TRR—Tire and wheel radial run-out. Not a cause of vibration below 20 mph (32 km/h). Speed required to cause vibration increases as run-out decreases. Automobile-speed-sensitive vibration.

WH—Wheel hop. Not a cause of vibration below 20 mph (32 km/h). Produces up-down movement in steering wheel and instrument panel along with mechanical vibration. Most noticeable between 20-40 mph (32-64 km/h). Caused by tires having radial run-out of more than 0.045 inch (1.14 millimeters). Do not attempt to correct by balancing; replace the tire. Automobile-speed-sensitive vibration.

TB—Tire balance. Static unbalance not a cause of vibration below 30 mph (48 km/h). Dynamic unbalance not a cause under 40 mph (64 km/h). Automobile- speed-sensitive vibration.

TLR—Tire and wheel lateral run-out. Not a cause of vibration below 55 mph (88 km/h) unless run-out is extreme. Automobile-speed-sensitive vibration.

TW—Tire wear. Abnormal wear can cause vibration in 30 to 55 mph (48 to 88 km/h) range and may also generate whine at

Correction Codes For Mechanical Vibrations Within Specific mph Ranges

Vibration Sensitivity	16	32	48	64	80	97	113	129	145
Car Speed Sensitive		←TRR———————————————————————————————→							
	←W→		←———TB——————————————————→						
		←———WH———→			←———TLR———→				
			←————PSY—————————————→						
		←——UJAN and TEB——→							
					←WB→				
Torque Sensitive	←————————UJAN and TEB————————————————————→								
	←UJA→						←UJA→		
Engine Speed Sensitive		←EA→							
	←DEM→								

Correction Codes For Audible Vibrations Within Specific mph Ranges

Vibration Sensitivity	16	32	48	64	80	97	113	129	145
Car Speed Sensitive		←———UJA———→			←—PSY—→				
		←UJ and WH and TEB→							
			←——————TW——————————————————→						
	←——————WB——————————————————————————→								
Torque Sensitive			←———————AN———————————→						
			←UJ and TEB→						
Engine Speed Sensitive	←————————————EA————————————————————→								
		←——ADB——→							
	←DEM→								

Figure 5-24. *(Courtesy of American Motors Corporation)*

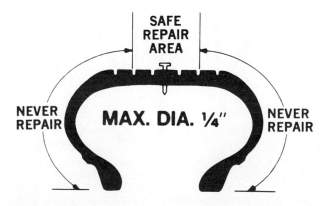

Figure 5-25. The area of safe tire repair is shown here.

SAFE REPAIR AREA

NEVER REPAIR   MAX. DIA. ¼″   NEVER REPAIR

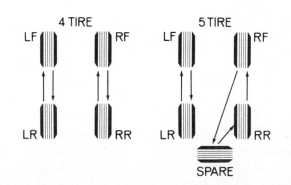

Figure 5-26. Proper sequence of tire rotation for radial tires. Tires should be rotated at 5,000-mile (8,000-kilometer) intervals or as recommended by manufacturer.

high-speed changing to growl at low speed. Automobile-speed-sensitive vibration.

W—Radial tire waddle. Normal condition with radial-ply tires. Unique construction causes side-to-side waddle motion at speeds up to 15 mph (24 km/h). Rotating tires front-to-rear may reduce condition. Replace tires if the condition cannot be reduced satisfactorily. Automobile-speed-sensitive vibration.

UJA—Universal joint angles. Incorrect angles may cause mechanical vibration below 20 mph (32 km/h) and mechanical or audible vibration at 35 to 55 mph (56 to 88 km/h). Torque-sensitive vibration.

UJ—Universal joints. If the ends of bearing crosses or bearing cups are galled, worn excessively, brinelled, or binding due to over-tightened U-bolts or clamp strap bolts, they will cause vibration at any speed. Torque-sensitive vibration.

PSY—Propeller shaft and yokes. Not a cause of vibration below 35 mph (56 km/h). Excessive run-out, unbalance, loss of balance weights, or undercoating on shaft will cause vibration at 35 mph (56 km/h) and above. Torque-sensitive vibration.

WB—Front wheel bearings. If loose, can cause automobile speed sensitive mechanical vibration at 35 mph (56 km/h) and above. If rough or damaged, can cause growl and grind noise at low speed or whine at high speed. Automobile-speed-sensitive vibration.

AN—Rear axle noise. Not a cause of vibration unless axle shaft is bent or shaft bearing has broken. Worn or damaged gears or bearings will cause noise in varying speed ranges in relation to amount of torque applied.

TEB—Transmission extension housing bushing. If worn or loose, can cause torque-sensitive mechanical vibration and oil leakage.

EA—Engine-driven accessories. Loose or broken air-conditioning compressor, power-steering pump, air pump, alternator, water pump, etc., can cause engine-speed-sensitive mechanical vibration. Usually apparent when transmission is placed in neutral and engine speed is increased.

ADB—Accessory drive belts. If excessively worn or loose, can cause engine-speed-sensitive audible vibration that sounds like droning, flutter, or rumbling noise.

DEM—Damaged engine mounts. If worn or broken, may allow engine or accessories to contact body, causing noise and vibration.

*Courtesy of American Motors Corp.

General Precautions

• Avoid getting grease, solvent, brake fluid, or dirty fingerprints on brake discs, pads, linings, and drums.

• Always use safe jacking and vehicle support procedures.

• Always tighten all fasteners to specified torque.

• Do not allow inflated tires to drop; they will bounce.

• Follow all manufacturer's procedures and specifications.

• Do not exceed 275 kPa (40 psi) pressure when initially inflating any tire, including compact spares. If 275 kPa (40 psi) pressure will not seat beads, deflate, relubricate and reinflate. Overinflation may cause the bead to break and cause serious personal injury.

• Test the repair with water or soap suds.

• Never try to convert a tubeless tire to a tubed tire to remedy serious damage.

• Never try to repair tires worn below 1/16-inch (1.5 mm) tread depth.

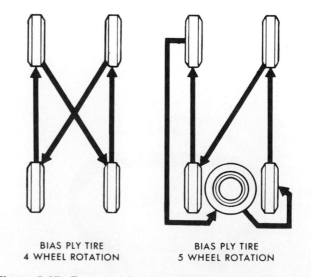

BIAS PLY TIRE
4 WHEEL ROTATION

BIAS PLY TIRE
5 WHEEL ROTATION

Figure 5-27. Four- and five-tire rotation sequence for bias-ply tires. *(Courtesy of General Motors Corporation)*

Wheel and Tire Service

Only small clean punctures should be repaired in tires. Repairs should be restricted to the safe repair area (Figure 5-23). Tire repair methods vary greatly and should be done according to the supplier's and vehicle manufacturer's procedures and recommendations.

Tire service life is greatly extended by periodic and systematic tire rotation. Follow the manufacturer's recommendations for best results. Wheel and tire radial and lateral runout must be within specifications for good handling and tire life. Wheel-and-tire assemblies must also be properly balanced.

Always replace valve stems when replacing tubeless tires. Be sure that the rims are clean for proper bead sealing.

Inspection and Cleaning

The condition of the wheels should be checked periodically. Replace any wheel that is bent, cracked, severely dented, or has excessive run-out. Also check the condition of the tire inflation valve. Replace the valve if worn, cracked, loose, or leaking air.

When cleaning steel or aluminum wheels, use a mild soap and water solution only and rinse with clean water. Do not use any type of caustic solution or abrasive substance, especially on forged aluminum wheels. After cleaning aluminum wheels, apply a coating of protective wax to preserve the finish and retain the original luster.

The finned urethane inserts on styled wheels may be cleaned using a sponge or soft bristle brush. Do not press overly hard on the inserts to clean them. They are flexible to a degree but can be damaged if due care is not exercised.

Always mount wheel assembly by tightening mounting nuts in proper sequence and to specified torque.

Tire Removal

Remove the valve core to deflate the tire completely. Unseat tire beads from the wheel, using bead breaker. Mount the wheel assembly in holding equipment for demounting. Follow the equipment manufacturer's directions; procedures vary depending on available equipment. Remove the tire from the wheel.

Puncture Repair Procedures

After removing the tire from the rim, probe repairable tire injuries in order to remove a nail or other damaging material. Make sure that the area around the injury is thoroughly dry. Scrape the damaged area with a sharp-edged tool and buff. Take care not to damage the liner or expose any cords.

Lubricate the injury by pushing the snout of the vulcanizing fluid can into the injury from both sides of the tire. Also pour vulcanizing fluid on the insertion tool and push it through with a twisting motion until it can be inserted and withdrawn easily.

Using a head-type or headless straight plug slightly larger than the size of the injury, place it in the eye of the insertion tool. When a headless straight plug is used, always back it up with a patch. Wet both the plug and insertion tool with vulcanizing fluid. Always pour directly from the can so as not to contaminate the can's contents.

While holding and stretching the long end of the plug, insert the plug into the injury from *inside* the tire. Hold and stretch the long end of the plug as it is forced into the injury until one end extends through it.

Remove the insertion and cut off the plug 1/16 inch above the surfaces. Do not pull on the plug while cutting. Do not wash previously prepared surface with solvent prior to application of vulcanizing fluid.

When using a cold patch, carefully remove the backing from the patch. Center the patch base over the damaged area on which vulcanizing fluid has been spread and allowed to dry. Roll the patch down firmly with the roller tool, working from the center out.

When using a hot patch, cover the buffed area with a light coat of cement of the type specified for the patch and allow it to dry. Remove the backing from the patch. Center the patch over injury. Clamp---finger tighten only. Apply heat, cure, and allow to cool. Before remounting the tire, clean and deburr the rim carefully.

Mounting Tires

If a rim is dirty or corroded or if the tire is not centered on the rim, the tire bead may "bind" on the rim and refuse to seat. Allowing

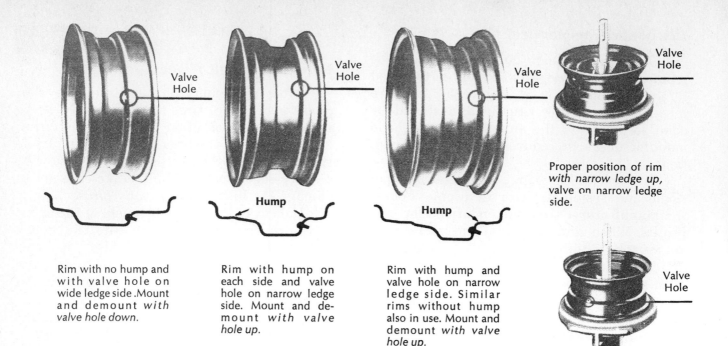

Valve Hole

Valve Hole

Valve Hole

Valve Hole

Proper position of rim *with narrow ledge up,* valve on narrow ledge side.

Hump

Hump

Rim with no hump and with valve hole on wide ledge side. Mount and demount *with valve hole down.*

Rim with hump on each side and valve hole on narrow ledge side. Mount and demount *with valve hole up.*

Rim with hump and valve hole on narrow ledge side. Similar rims without hump also in use. Mount and demount *with valve hole up.*

Valve Hole

Proper position of rim *with narrow ledge up,* valve on wide ledge side.

IMPORTANT: **In Both mounting and Demounting, Always Start with the Narrow Bead Ledge of Rim Up.**

Figure 5-28. Important wheel and tire service information that must be followed to service wheel-and-tire assemblies properly.

air pressure to build within the assembly in an attempt to seat the bead is a *dangerous practice*. Inflation beyond 40 psi (275 kPa) may break the bead (or even the rim) with explosive force. This can cause serious injury to the person inflating the tire. Injuries caused by such explosions include severed fingers, broken arms, broken jaws, and severe facial lacerations.

1. Be certain that rim flanges and bead ledge (especially hump and radius) areas are smooth and clean. Remove any oxidized rubber, dried soap solution, rust, heavy paint, etc., with a wire brush or a file.

2. Lubricate tire beads, rim flanges, and bead ledge areas with a liberal amount of thin vegetable oil soap solution or with an approved rubber lubricant. Start the mounting procedure with the narrow bead ledge of the rim up at all times.

3. Be sure that the assembly is securely locked down on the mounting machine.

4. *Use a tire mounting band.* The use of a tire mounting band (or bead expander) is helpful when inflating tubeless tires. This device constricts the tread center line of the

Figure 5-29. Several types of tubeless-tire valve stems. Stems come in different lengths and diameters for different-sized wheel stem holes.

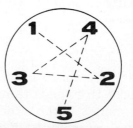

Figure 5-30. Always follow the above star pattern when tightening wheel bolts or lug nuts. This prevents misalignment and distortion.

192

tire, thereby helping to force the beads onto the bead seats of the rim. Follow these steps.

a. When the tire is on the wheel, and before inflating, attach the bead expander around the center of the tread.

b. Inflate the tire sufficiently (10 psi (69 kPa) or less) to move the tire beads out to contact bead seats of rim. Then, as a safety precaution, remove the expander. Never exceed 10 pounds pressure with the mounting band on the tire.

c. Increase air pressure, as needed, up to 40 psi (275 kPa) to seat the tire beads fully on the rim.

d. Check for leakage and, if none, adjust air pressure to recommended pressure. *Important: On safety or hump-type rims, make sure that tire beads have moved over the hump on the rim and are fully seated.*

5. Do not allow air pressure to exceed 40 psi (275 kPa) during the bead-seating process. If beads have not seated by the time pressure reaches 40 psi (275 kPa), deflate the assembly, reposition the tire on the rim, and relubricate and reinflate it to recommended operating pressure.

6. Make certain that the valve core is inserted in the valve stem. Worn valves should be replaced, using the valve designated by the manufacturer, since valves vary as to length

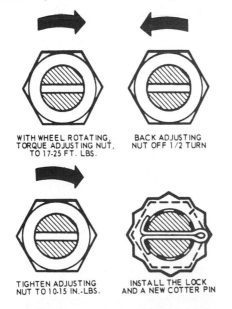

WITH WHEEL ROTATING, TORQUE ADJUSTING NUT, TO 17-25 FT. LBS.

BACK ADJUSTING NUT OFF 1/2 TURN

TIGHTEN ADJUSTING NUT TO 10-15 IN.-LBS.

INSTALL THE LOCK AND A NEW COTTER PIN

Figure 5-31. Example of wheel-bearing nut adjustment sequence and properly installed lock and cotter pin. Follow the manufacturer's specifications whenever servicing wheel bearings. *(Courtesy of Ford Motor Co. of Canada Ltd.)*

and diameter. Valve caps should be screwed on finger-tight.

7. Use an extension gauge with clip-on chuck so that air pressure buildup can be closely watched and so that you can stand well back from the assembly during the seating process.

PART 4 WHEEL-BALANCING PROCEDURE

PROCEDURE

• Clean all dirt and foreign matter from the wheels and remove all wheel weights.

• Clean larger stones from tire tread.

• Check and correct tire and wheel run-out in excess of specifications.

• Correct tire inflation pressure.

• For off-the-car balancing, mount the wheel as recommended.

• For on-the-car balancing, position jacks and wheel spinner safely and mount attachments as recommended.

• Stay clear of the plane of tire rotation when spinning tire and wheel assembly (flying stones can injure).

• Attach all balancing weights securely.

• If the wheel was removed for balancing, install and tighten to proper torque after balancing.

• Replace hub cap or wheel cover using a rubber mallet around the edges.

Balancing Drive Wheels on the Car

Standard Differential

• Jack up one side only and block the other wheel.

• *Caution:* The wheel speed is double that of the speedometer reading. Do not overspeed; damage to the differential may result. Maximum speed is 35 mph (60 km/h).

• Prepare the wheel for balancing as indicated and balance the wheel assembly.

• Repeat the procedure for the other side.

Limited Slip Differential

• Jack up both rear wheels and support with jack stands; block the front wheels.

• Remove one rear wheel.

• Prepare the other wheel (on the car) for balancing as indicated and proceed to balance.

• The speedometer reading is actual wheel speed on limited slip differential vehicles when balancing.

• Follow the manufacturer's recommendations for procedures and speeds.

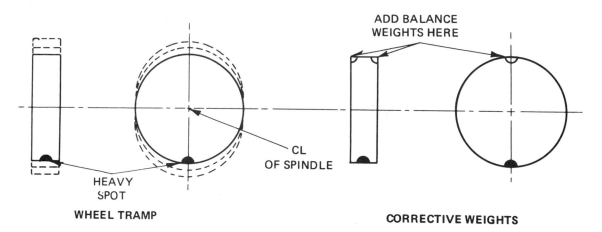

Figure 5-32. A wheel assembly must be statically in balance to prevent wheel tramp. A wheel that is not statically in balance has a heavy spot. To bring the wheel into static balance, place two equal-sized weights 180° from the heavy spot, one on each side of the wheel rim. The combined weight of the two balance weights must equal the weight of the heavy spot. A statically balanced wheel has equal weight distribution radially around the axis of rotation. *(Courtesy of General Motors Corporation)*

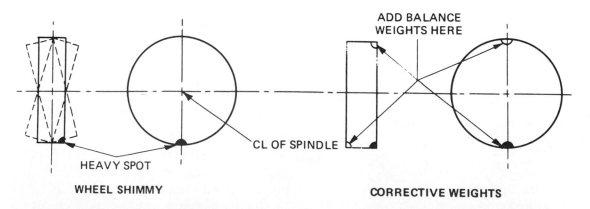

Figure 5-33. A wheel must also be in dynamic balance to prevent wheel shimmy. A wheel that is not in dynamic balance has a heavy spot to one side of the center of the plane of tire rotation. To bring the wheel into dynamic balance, the combined weight of the two balance weights placed as above must equal the heavy spot. A dynamically balanced wheel has equal weight on each side of the wheel center line or plane of rotation. *(Courtesy of General Motors Corporation)*

PROJECT NO. 5-A

TITLE: <u>WHEELS AND TIRES</u>

STUDENT'S NAME _____

PERFORMANCE OBJECTIVES

After sufficient opportunity to study this portion of the text and the appropriate training models and with the instructor's supervision and demonstrations, you should be able to perform the following tasks at the request of your instructor.

TASK 1---State the purpose of the tire-and-wheel assemblies.

TASK 2---Describe the construction and operation of the wheels and tires under normal and abnormal operating conditions.

TASK 3---Describe the effects of such operation on the wheels and tires.

PERFORMANCE EVALUATION

Your instructor may require you to perform these tasks in any of the following ways in order to evaluate your performance:
- By asking test questions
- By asking you to describe the performance of these tasks in writing
- By asking you to describe the performance of these tasks orally

WHEELS AND TIRES

**TITLE: <u>WHEEL AND
TIRE DIAGNOSIS
AND SERVICE</u>**

STUDENT'S NAME _____

PERFORMANCE OBJECTIVES

After sufficient opportunity to study this portion of the text and the appropriate training models and with the instructor's supervision and demonstrations, you should be able to perform the following tasks at the request of your instructor.

TASK 1---Complete the diagnostic chart provided. Diagnose wheel and tire problems as specified by your instructor.

TASK 2---Perform tire service according to procedures given in the text, in the shop manual, and by your instructor. Use the worksheet provided.

PERFORMANCE EVALUATION

Your ability to perform these tasks will be evaluated by your instructor on the basis of how accurately you have followed the procedures and specifications given to you by your instructor, by the text, and by the manufacturer's service manual.

PROBLEM	CAUSE	CORRECTION
Front End Noise	1. 2.	1. 2.
Too much free play in steering	1. 2.	1. 2.
Front-wheel shimmy.	1. 2. 3.	1. 2. 3.
Instability.	1. 2.	1. 2.
Tires wear unevenly.	1. 2. 3.	1. 2. 3.

— TIRE SIZE   L.F. _____   R.F. _____

                               L.R. _____   R.R. _____

— TIRE MAKE AND TYPE   L.F. _____

                               R.F. _____

                               L.R. _____

                               R.R. _____

— TREAD WEAR   OK _____   NOT OK _____   EXPLAIN _____

— TIRE BODY CONDITION   OK _____   NOT OK _____   EXPLAIN _____

— VALVE STEMS   OK _____   NOT OK _____   EXPLAIN _____

— TIRE REPAIR REQUIRED?   EXPLAIN _____

— WHEEL CONDITION   OK _____   NOT OK _____   EXPLAIN _____

— WHEEL BEARINGS   OK _____   NOT OK _____   EXPLAIN _____

— WHEEL BEARINGS REPACKED?   YES _____   NO _____

— WHEEL BEARING ADJUSTMENT PROCEDURE AND SPECIFICATIONS _____

_____

— WHEEL NUT TORQUE   _____

— TIRES ROTATED?   YES _____   NO _____

— ROTATION PATTERN USED   _____

STUDENT SIGN HERE _____

INSTRUCTOR VERIFY _____

Self-Check

1. What three materials are used for wheel construction?
2. How is wheel size determined or designated?
3. What is the purpose of the wheel rim drop center?
4. Why are tire beads required?
5. List three types of tire construction.
6. What is the purpose of tire tread sipes?
7. Define the slip angle of a tire.
8. What causes radial tire lead?
9. How is tire load rating designated?
10. How should wheel and tire radial and lateral run-out be checked?
11. Do not exceed_____air pressure when inflating an automobile tire.
12. Tire repairs should not be attempted on sidewalls. True or false?
13. What is the purpose of a tire-mounting band?
14. Define static wheel balance.

15. Define dynamic wheel balance.
16. How should a wheel and tire be prepared for balancing?

Performance Evaluation

After through study of this chapter and sufficient practical work on the appropriate components, and with the appropriate shop manual, tools, and equipment, you should be able to do the following.

1. Follow the accepted general precautions while servicing wheels and tires.

2. Correctly diagnose specified wheel and tire problems.

3. Remove, repair, balance, and install wheel-and-tire assemblies.

4. Properly prepare the vehicle for customer acceptance.

5. Complete the Self-Check with at least 80 percent accuracy.

6. Complete all practical work with 100 percent accuracy.

Chapter 6

Steering Systems

The steering system is the means by which the driver of a vehicle is able to control the position of the front wheels. The system must provide ease of handling, good directional control, and stability. This is achieved by the steering system in conjunction with the suspension system.

PART 1   STEERING LINKAGE

The steering linkage includes the steering arms, which are attached to (or are a part of) the steering knuckle, tie rods and tie rod ends, adjusting sleeves and clamps, a center link (relay rod or drag link), an idler arm, and a pitman arm. The arrangement of these parts is shown in Figure 6-2. Another type of linkage is shown in Figure 6-3. These are the two most common types of steering linkage on passenger cars.

Other types of linkage arrangements are used on older cars and on trucks. The wheel assembly, including the steering knuckle, pivots at the ball joints, allowing the front wheels to turn to the right or to the left. The steering arms being attached to the steering knuckle turn the wheels to right or left as the linkage moves from side to side.

The pivoting action in the linkage is provided by the tie rod ends. The outer ends of the tie rods are attached to the steering arms and the inner ends to the center link. Threaded adjusting sleeves between the outer tie rod end and tie rod connect the two and provide a means of adjusting tie rod length. Adjusting sleeves are split for ease of adjustment and are clamped at each end. The clamps provide a means of preventing the sleeves from turning on the rods during vehicle operation. The inner ends of the two tie rods are attached to the center link by means of tie rod ends.

The center link is supported at one end by the idler arm and at the other by the pitman arm. The pitman arm swings through an arc from side to side, as does the idler arm, when the steering wheel is turned from right to left and back.

The idler arm is bolted to the frame on the right side of left-hand-drive vehicles. The pitman arm is attached to the splined sector shaft of the steering gear with a nut and lock washer.

On rack-and-pinion steering two tie rods are used, one on each side. These rods are threaded at both ends. The inner end is attached to the rack and the outer end to the steering arm. Lock nuts hold the adjustment in place.

PART 2   MANUAL-STEERING GEARS

The steering gear provides the means of converting the turning of the steering wheel to side-to-side movement of the steering linkage. On some vehicles the steering gear converts the turning of the steering wheel to fore-and-aft movement of the pitman arm, which is then converted to side-to-side linkage movement by a specially shaped steering arm. The steering gear is bolted to the frame. The most common types of manual-steering gears are the recirculating-ball type and the rack-and-pinion type.

Turning the steering wheel on a vehicle equipped with a recirculating-ball type of steering gear turns the grooved worm shaft in the gear. Turning the grooved worm shaft causes the ball nut to thread its way up or down the worm. The teeth of the ball nut push against the sector shaft teeth as the worm

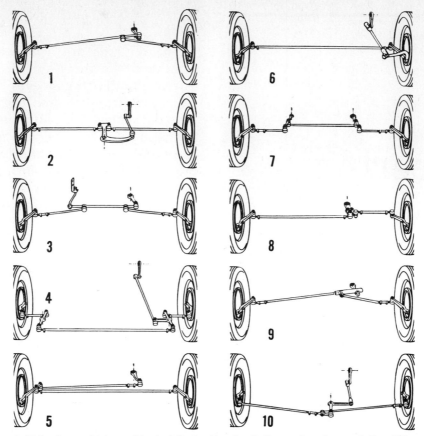

1. Haltenberger Linkage (Vertical Socket Type)   2. Center Arm Steer   3. Parallelogram Linkage (Solid Center Link)   4. Parallelogram Linkage (Center Link Ahead of Axle)   5. Cross Steer   6. Fore-Aft and Cross Steer   7. Parallelogram Steer   8. Long-Short Arm Linkage   9. Haltenberger Linkage   10 Haltenberger Linkage (Ahead of Axle)

Figure 6-1. Steering linkage design variations.

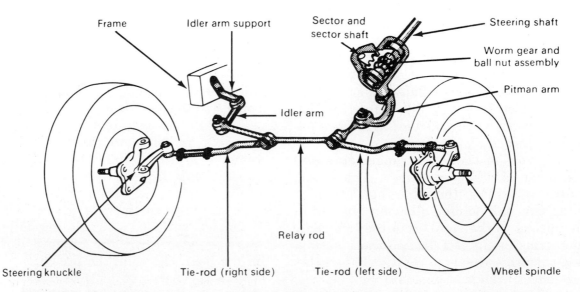

Frame

Idler arm support

Sector and sector shaft

Steering shaft

Worm gear and ball nut assembly

Pitman arm

Idler arm

Steering knuckle

Tie-rod (right side)

Relay rod

Tie-rod (left side)

Wheel spindle

Figure 6-2. Manual or integral power-steering linkage components. This type of linkage is known as parallelogram linkage. This illustration shows a manual type of steering gear.

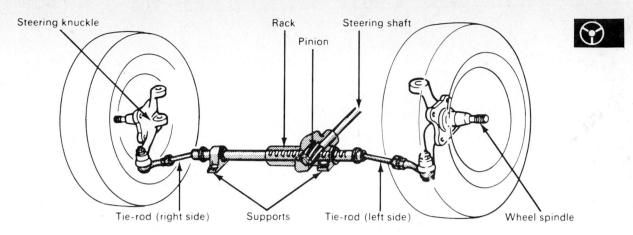

Figure 6-3. Rack-and-pinion type of steering linkage. Note that the steering knuckle arms are pointing forward instead of rearward as in Figure 6-2.

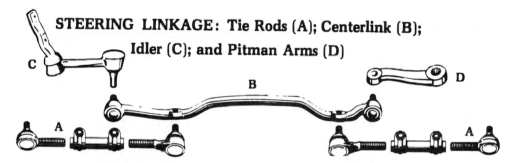

STEERING LINKAGE: Tie Rods (A); Centerlink (B); Idler (C); and Pitman Arms (D)

Figure 6-4. Manual-steering linkage components, disassembled view. Some tie rod ends are threaded into other linkage parts, while others are part of the linkage itself and must be replaced as an assembly. The center link (B) is an example of this.

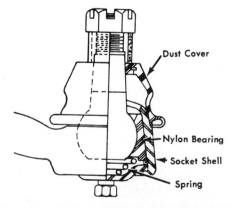

Figure 6-5. Tie rod end cross-sectional view showing spring-loaded ball and socket. Note dust boot and plug for grease fitting. Tapered stud fits into tapered hole of linkage and is fastened with castellated nut and cotter pin.

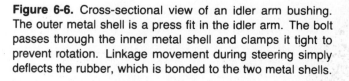

Figure 6-6. Cross-sectional view of an idler arm bushing. The outer metal shell is a press fit in the idler arm. The bolt passes through the inner metal shell and clamps it tight to prevent rotation. Linkage movement during steering simply deflects the rubber, which is bonded to the two metal shells.

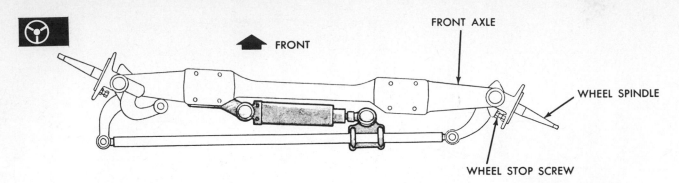

FRONT AXLE

FRONT

WHEEL SPINDLE

WHEEL STOP SCREW

Figure 6-7. Steering linkage damper controls wheel shimmy. Operation is similar to a shock absorber. The moveable piston rod is connected to the steering linkage while the cylinder is anchored at the axle in this illustration or to the frame in other applications. *(Courtesy of Crysler Corporation)*

shaft is turned. The sector shaft teeth and sector shaft are integral. As the sector shaft teeth move back and forth, the sector shaft is forced to turn back and forth as well. This causes the pitman arm to move from side to side to actuate the linkage.

The steering gear also provides a gear reduction. This is needed to reduce the amount of effort required to turn the steering wheel, particularly when parking. During straight-ahead driving this also reduces the possibility of oversteering. This gear reduction is known as steering gear ratio. Ratios vary considerably, depending largely on vehicle size and weight. Larger, heavier vehicles require a greater reduction in manual steering gears. The steering gear also absorbs much of the road shock, which attempts to turn the front wheels right or left.

The rack-and-pinion type of steering mechanism is used on many smaller cars. A small toothed gear is connected to the bottom of the steering shaft through a flexible coupling or shaft. The toothed gear or pinion is in mesh with teeth on a long horizontal bar or rack with teeth on one side. As the steering wheel is turned, the pinion turns, causing the rack to move from side to side. The ends of the rack are connected to the tie rods, which in turn are connected to the steering arms at the wheels. The rack-and-pinion assembly is enclosed in a housing and sealed at each end with a flexible rubber boot. It is also sealed at the pinion shaft. The entire assembly is attached to the vehicle frame or fire wall.

PART 3 POWER STEERING

The power-steering unit is designed to reduce the amount of effort required to turn the steering wheel. It also reduces driver fatigue and increases safety by providing better control on rough road surfaces or during tire failure.

The system includes a pump driven by a belt from the crankshaft pulley. This provides the fluid pressure and flow needed to operate the system. Maximum pressure is controlled by a pressure-regulating valve in the pump. A pressure line and a return line connect the pump to the system. Three types of power-steering pumps are used. All operate similarly and differ mainly in rotor design. Some pumps have rollers; others have slippers or vanes.

The integral type of power-steering unit combines the power piston and cylinder with the steering gear and flow control valve all in a single assembly. Typical integral power-steering gear components are shown in Figure 6-14. System operation is shown and described in Figures 6-16 to 6-20.

Another type of power-steering system is the linkage booster type. Major components of this system are shown in Figures 6-23 and 6-24. System operation is shown in Figure 6-25.

Rack-and-pinion power steering incorporates the control valve, power piston, and gear in a single assembly.

Power steering provides most of the force needed for steering. The remaining steering effort provides needed driver "feel" for good steering control.

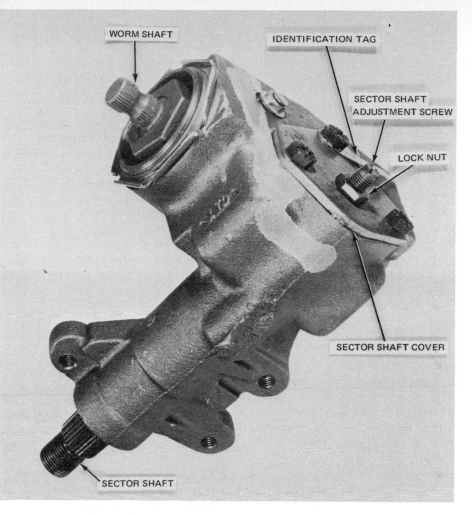

Figure 6-8. Outside view of recirculating ball type of manual steering gear. *(Courtesy of Ford Motor Co. of Canada Ltd.)*

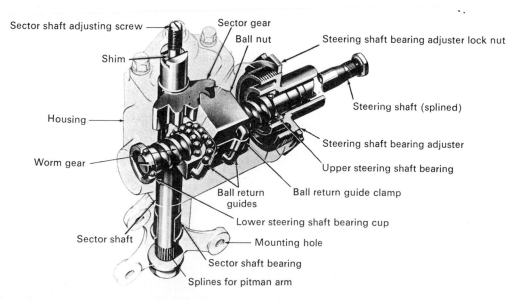

Figure 6-9. Phantom view of recirculating ball type of manual-steering gear showing names of major parts. Steering effort is reduced by using the worm-and-recirculating ball-nut design. *(Courtesy of Ford Motor Co. of Canada Ltd.)*

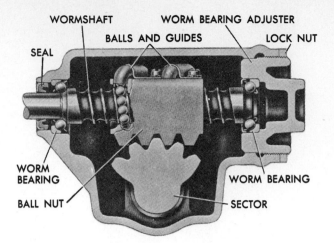

Figure 6-10. Steering gear ball nut and sector teeth shown in mesh. Note worm bearings supporting worm shaft and worm bearing adjuster and lock nut. There are two separate ball circuits. Balls do not pass from one guide to the other. *(Courtesy of Ford Motor Co. of Canada Ltd.)*

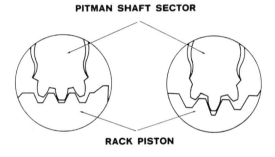

Figure 6-11. Constant-ratio and variable-ratio steering gear pitman shaft and rack piston teeth compared. The ratio of front-wheel turn to steering wheel movement remains constant throughout a complete turn right to left or left to right with the constant-ratio steering gear. The variable-ratio steering gear provides relatively little front-wheel movement during the first quarter of a turn to right or left from straight ahead. Front-wheel movement increases at a faster rate as the steering wheel is turned farther. The advantage is quicker maneuverability for parking. *(Courtesy of General Motors Corporation)*

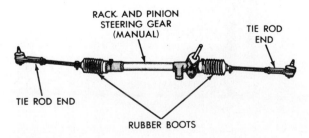

Figure 6-12. Rack-and-pinion steering gear assembly incorporates gear and linkage as a unit. *(Courtesy of Chrysler Corporation)*

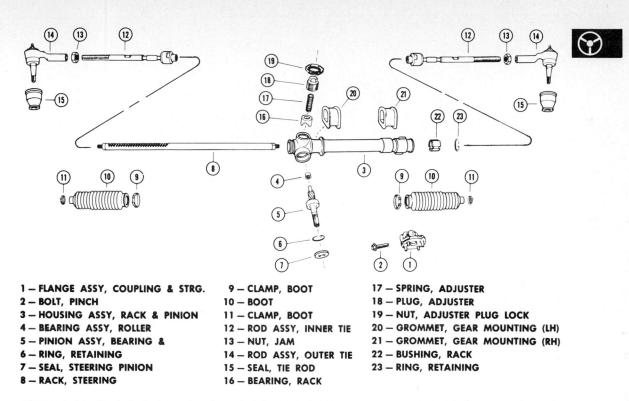

1 — FLANGE ASSY, COUPLING & STRG.
2 — BOLT, PINCH
3 — HOUSING ASSY, RACK & PINION
4 — BEARING ASSY, ROLLER
5 — PINION ASSY, BEARING &
6 — RING, RETAINING
7 — SEAL, STEERING PINION
8 — RACK, STEERING

9 — CLAMP, BOOT
10 — BOOT
11 — CLAMP, BOOT
12 — ROD ASSY, INNER TIE
13 — NUT, JAM
14 — ROD ASSY, OUTER TIE
15 — SEAL, TIE ROD
16 — BEARING, RACK

17 — SPRING, ADJUSTER
18 — PLUG, ADJUSTER
19 — NUT, ADJUSTER PLUG LOCK
20 — GROMMET, GEAR MOUNTING (LH)
21 — GROMMET, GEAR MOUNTING (RH)
22 — BUSHING, RACK
23 — RING, RETAINING

Figure 6-13. Exploded view of rack-and-pinion steering gear components. *(Courtesy of General Motors Corporation)*

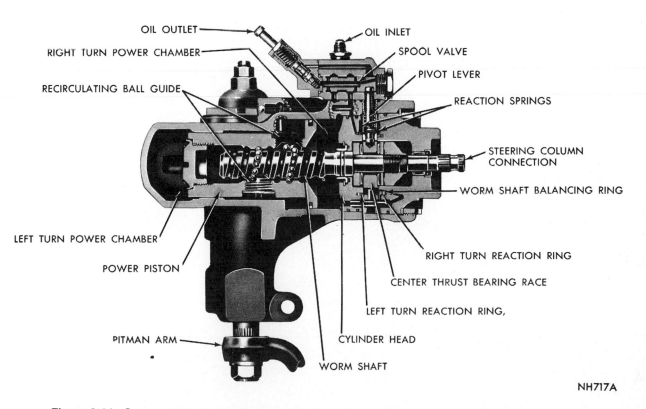

NH717A

Figure 6-14. Cross-sectional view of spool-valve type of integral power steering gear showing major parts. *(Courtesy of Chrysler Corporation)*

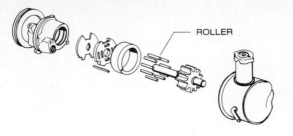

ROLLER

VANE TYPE

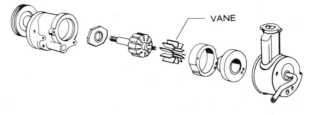

VANE

SLIPPER TYPE

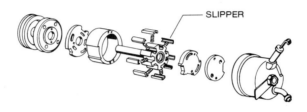

SLIPPER

Figure 6-15(A). Common power-steering pump designs. Operation of these pumps is similar even though the pumping elements (vanes, rollers, slippers) differ. *(Courtesy of Moog Automotive Inc.)*

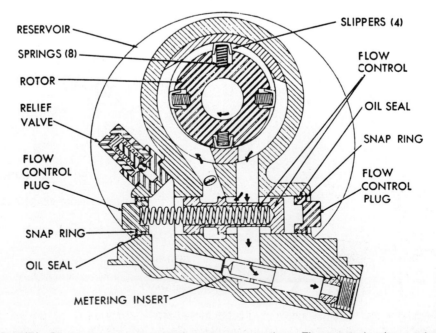

RESERVOIR

SPRINGS (8)

ROTOR

RELIEF VALVE

FLOW CONTROL PLUG

SNAP RING

OIL SEAL

METERING INSERT

SLIPPERS (4)

FLOW CONTROL

OIL SEAL

SNAP RING

FLOW CONTROL PLUG

Figure 6-15(B). Slipper-type power-steering pump operation. Flow-control valve controls flow to steering gear. Relief valve limits maximum pump pressure. *(Courtesy of Applied Power)*

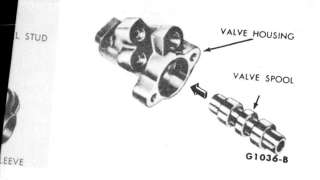

VALVE HOUSING

VALVE SPOOL

G1036-B

booster power-steering control valve assembly.

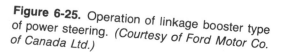

Figure 6-25. Operation of linkage booster type of power steering. *(Courtesy of Ford Motor Co. of Canada Ltd.)*

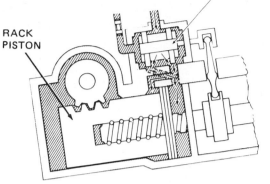

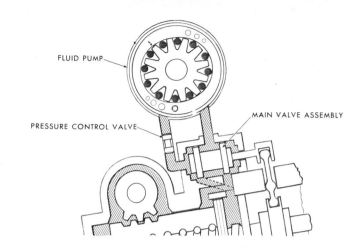

FLUID PUMP

PRESSURE CONTROL VALVE

MAIN VALVE ASSEMBLY

Figure 6-16. The power-steering pump (fluid pump) provides the hydraulic pressure for the power steering gear. Pressure is regulated by the pressure control valve and is directed to the rack piston by the main valve assembly. This provides power-assisted steering whenever the steering wheel is turned. *(Courtesy of Chrysler Corporation)*

FLUID CIRCULATES THROUGH VALVE

RACK PISTON

Figure 6-17. During straight-ahead driving, hydraulic pressure is directed equally to both sides of the rack piston. *(Courtesy of Chrysler Corporation)*

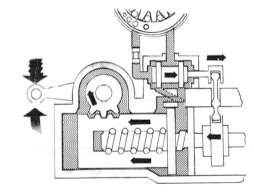

Figure 6-18. If a front wheel hits a bump or hole in the road, the front wheels are reflected to the right or left. This creates a tendency to self-steer to right or left. This action is transferred to the control valve as shown by arrows causing hydraulic pressure to offset the self-steering tendency. *(Courtesy of Chrysler Corporation)*

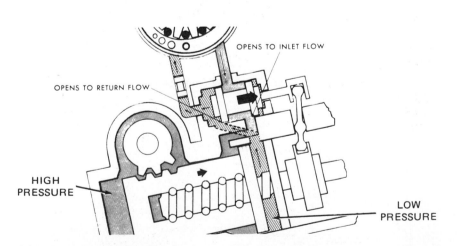

OPENS TO INLET FLOW

OPENS TO RETURN FLOW

HIGH PRESSURE

LOW PRESSURE

Figure 6-19. A left turn causes the worm shaft to thread into the rack piston, which pulls the piston up. A reaction on the pivot moves the control valve up, which directs hydraulic pressure to the bottom side of the rack piston. At the same time a return passage is opened from the chamber above the piston to allow fluid return and a pressure drop in this chamber. In effect, the pressure difference on the two sides of the piston assists the piston to move up which provides steering assist. *(Courtesy of Chrysler Corporation)*

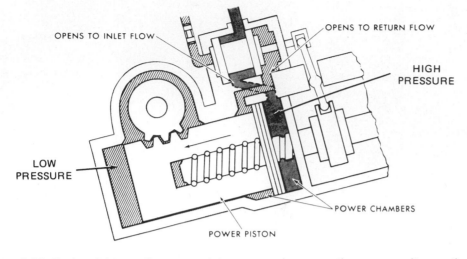

OPENS TO INLET FLOW

OPENS TO RETURN FLOW

HIGH
PRESSURE

LOW
PRESSURE

POWER CHAMBERS

POWER PISTON

Figure 6-20. During right turn the power piston moves down, creating an opposite reaction on the control valve as compared to Figure 6-19. The higher pressure above the piston helps move the piston down to the low-pressure area, providing steering assist. *(Courtesy of Chrysler Corporation)*

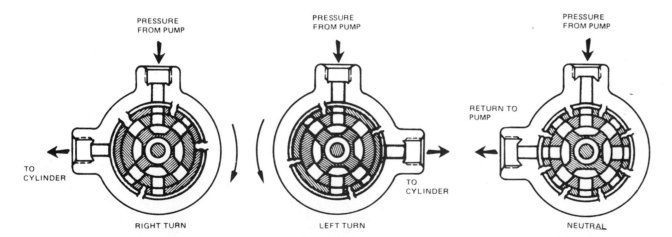

PRESSURE
FROM PUMP

PRESSURE
FROM PUMP

PRESSURE
FROM PUMP

RETURN TO
PUMP

TO
CYLINDER

TO
CYLINDER

RIGHT TURN

LEFT TURN

NEUTRAL

Figure 6-21. Hydraulic pressure flow paths of rotary-type control valve. *(Courtesy of Ford Motor Co. of Canada Ltd.)*

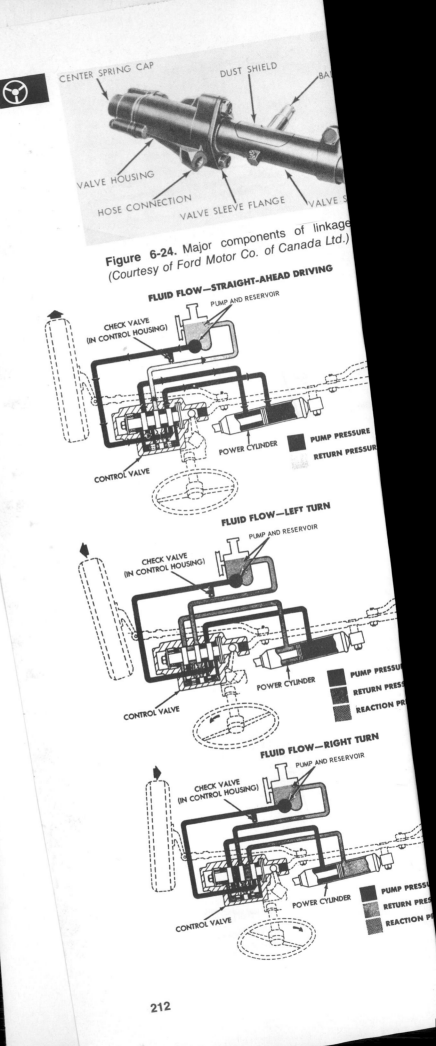

CENTER SPRING CAP

DUST SHIELD

BA

VALVE HOUSING

HOSE CONNECTION

VALVE SLEEVE FLANGE

VALVE S

Figure 6-24. Major components of linkage
(Courtesy of Ford Motor Co. of Canada Ltd.)

FLUID FLOW—STRAIGHT-AHEAD DRIVING

PUMP AND RESERVOIR

CHECK VALVE
(IN CONTROL HOUSING)

POWER CYLINDER

PUMP PRESSURE

RETURN PRESSUR

CONTROL VALVE

FLUID FLOW—LEFT TURN

PUMP AND RESERVOIR

CHECK VALVE
(IN CONTROL HOUSING)

POWER CYLINDER

PUMP PRESSU

RETURN PRES

REACTION PR

CONTROL VALVE

FLUID FLOW—RIGHT TURN

PUMP AND RESERVOIR

CHECK VALVE
(IN CONTROL HOUSING)

POWER CYLINDER

PUMP PRESS

RETURN PRE

REACTION P

CONTROL VALVE

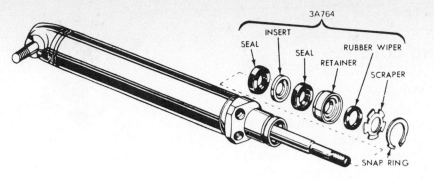

Figure 6-26. Power cylinder for linkage booster power steering. The piston rod (right) is attached to the vehicle frame and the other end of the cylinder is attached to the center link of the steering linkage. *(Courtesy of Ford Motor Co. of Canada Ltd.)*

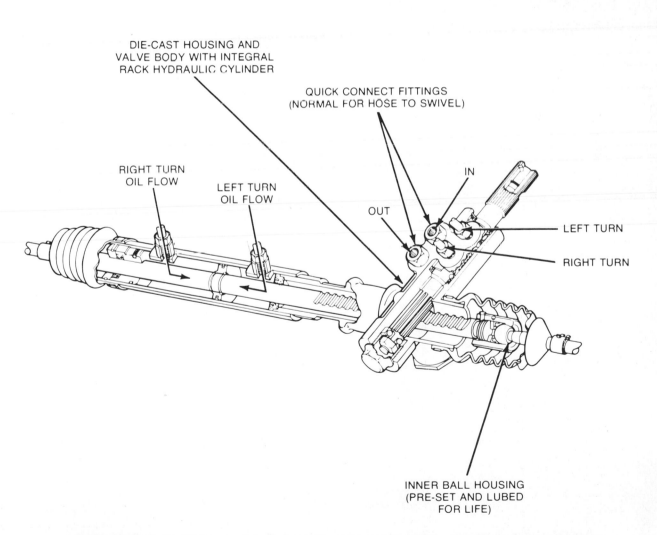

Figure 6-27. Typical power rack-and-pinion steering gear operation. *(Courtesy of Ford Motor Co. of Canada Ltd.)*

PART 4 STEERING COLUMNS

The steering column connects the steering gear to the steering wheel by means of a shaft and one or more flexible couplings. The steering column shaft is enclosed in a tubular mast jacket that is collapsible in case of impact due to collision. The steering column shaft and gear shift control rod are also collapsible. The steering column assembly includes such items as the turn signal switch, headlight dimmer switch, windshield wiper switch, ignition switch, steering wheel lock, gear shift lever and rod, speed control switch, horn switch, and necessary electrical wiring. Brackets at the dashboard and the floor panel hold the assembly in position. Some steering columns can be tilted or telescoped to adjust the steering wheel position to the driver's preference.

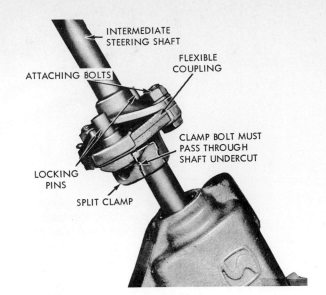

Figure 6-28. Details of flexible coupling between steering column shaft and steering gear shaft. A flexible coupling is required since the steering column shaft and steering gear shaft are not in alignment with each other. *(Courtesy of General Motors Corporation)*

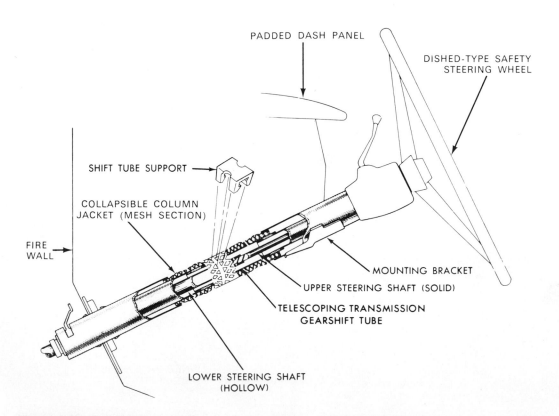

Figure 6-29. Collapsible steering column components. Collapsible steering columns have been a major safety feature for many years. *(Courtesy of Chrysler Corporation)*

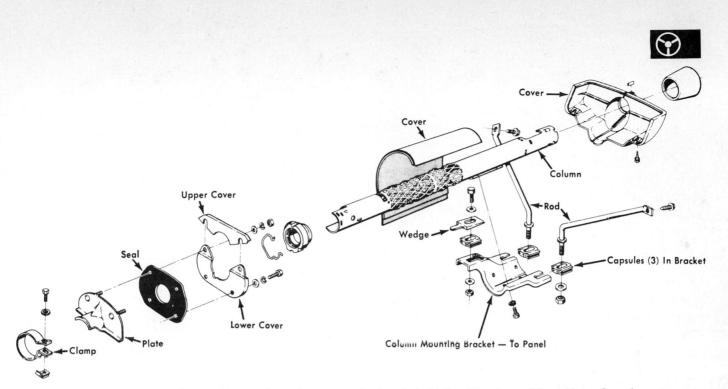

Figure 6-30. Collapsible steering column mounting bracket details. *(Courtesy of Ford Motor Co. of Canada Ltd.)*

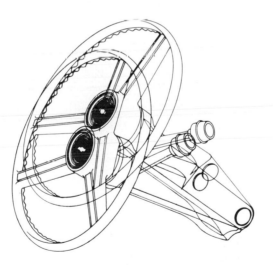

Figure 6-31. Some steering columns are designed to allow several vertical positions. Various tilt and telescope positions can be achieved to suit the driver's preference.

ON POSITION

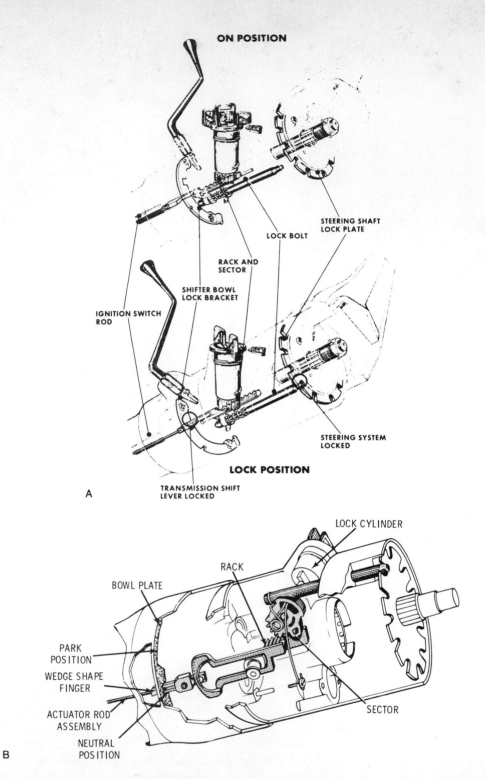

STEERING SHAFT
LOCK PLATE

LOCK BOLT

RACK AND
SECTOR

SHIFTER BOWL
LOCK BRACKET

IGNITION SWITCH
ROD

STEERING SYSTEM
LOCKED

LOCK POSITION

TRANSMISSION SHIFT
LEVER LOCKED

A

LOCK CYLINDER

RACK

BOWL PLATE

PARK
POSITION

WEDGE SHAPE
FINGER

ACTUATOR ROD
ASSEMBLY

SECTOR

NEUTRAL
POSITION

B

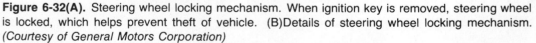

Figure 6-32(A). Steering wheel locking mechanism. When ignition key is removed, steering wheel is locked, which helps prevent theft of vehicle. (B)Details of steering wheel locking mechanism. *(Courtesy of General Motors Corporation)*

DIAGNOSTIC CHART

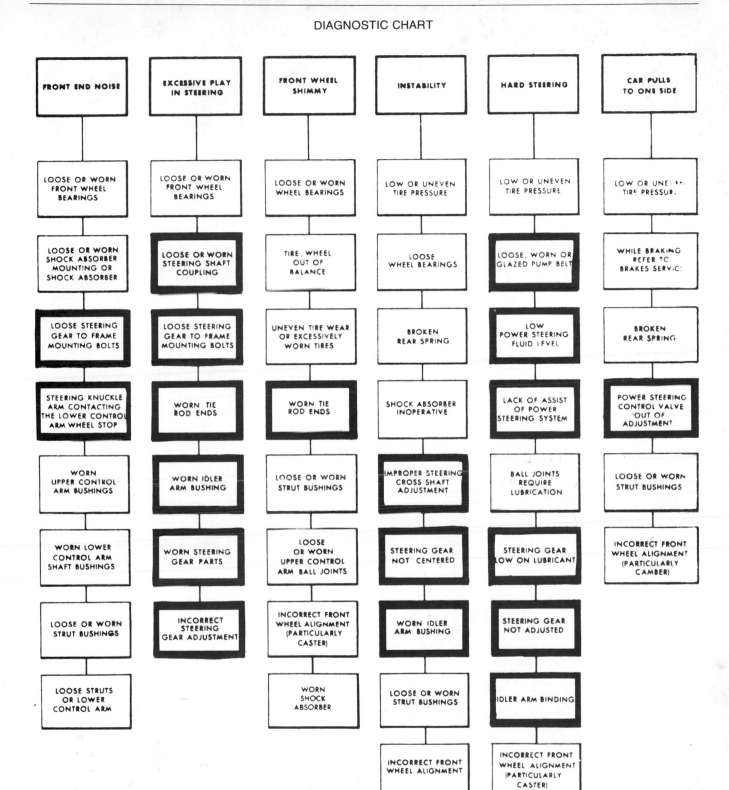

FRONT END NOISE	EXCESSIVE PLAY IN STEERING	FRONT WHEEL SHIMMY	INSTABILITY	HARD STEERING	CAR PULLS TO ONE SIDE
LOOSE OR WORN FRONT WHEEL BEARINGS	LOOSE OR WORN FRONT WHEEL BEARINGS	LOOSE OR WORN WHEEL BEARINGS	LOW OR UNEVEN TIRE PRESSURE	LOW OR UNEVEN TIRE PRESSURE	LOW OR UNEVEN TIRE PRESSURE
LOOSE OR WORN SHOCK ABSORBER MOUNTING OR SHOCK ABSORBER	LOOSE OR WORN STEERING SHAFT COUPLING	TIRE, WHEEL OUT OF BALANCE	LOOSE WHEEL BEARINGS	LOOSE, WORN OR GLAZED PUMP BELT	WHILE BRAKING REFER TO BRAKES SERVICE
LOOSE STEERING GEAR TO FRAME MOUNTING BOLTS	LOOSE STEERING GEAR TO FRAME MOUNTING BOLTS	UNEVEN TIRE WEAR OR EXCESSIVELY WORN TIRES	BROKEN REAR SPRING	LOW POWER STEERING FLUID LEVEL	BROKEN REAR SPRING
STEERING KNUCKLE ARM CONTACTING THE LOWER CONTROL ARM WHEEL STOP	WORN TIE ROD ENDS	WORN TIE ROD ENDS	SHOCK ABSORBER INOPERATIVE	LACK OF ASSIST OF POWER STEERING SYSTEM	POWER STEERING CONTROL VALVE OUT OF ADJUSTMENT
WORN UPPER CONTROL ARM BUSHINGS	WORN IDLER ARM BUSHING	LOOSE OR WORN STRUT BUSHINGS	IMPROPER STEERING CROSS SHAFT ADJUSTMENT	BALL JOINTS REQUIRE LUBRICATION	LOOSE OR WORN STRUT BUSHINGS
WORN LOWER CONTROL ARM SHAFT BUSHINGS	WORN STEERING GEAR PARTS	LOOSE OR WORN UPPER CONTROL ARM BALL JOINTS	STEERING GEAR NOT CENTERED	STEERING GEAR LOW ON LUBRICANT	INCORRECT FRONT WHEEL ALIGNMENT (PARTICULARLY CAMBER)
LOOSE OR WORN STRUT BUSHINGS	INCORRECT STEERING GEAR ADJUSTMENT	INCORRECT FRONT WHEEL ALIGNMENT (PARTICULARLY CASTER)	WORN IDLER ARM BUSHING	STEERING GEAR NOT ADJUSTED	
LOOSE STRUTS OR LOWER CONTROL ARM		WORN SHOCK ABSORBER	LOOSE OR WORN STRUT BUSHINGS	IDLER ARM BINDING	
			INCORRECT FRONT WHEEL ALIGNMENT	INCORRECT FRONT WHEEL ALIGNMENT (PARTICULARLY CASTER)	

General Precautions

When servicing steering systems or suspension systems, safe vehicle operation should always be ensured. Follow the manufacturer's specifications and recommendations. Some additional points to remember follow.

- Perform only quality workmanship.
- Always support the vehicle safely when working under the vehicle.
- Use correct equipment and tools in good condition, and use them as recommended.
- Do not use heat on parts that are to be reused.
- Replace all damaged, bent, or worn parts.
- Always use a torque wrench to tighten fasteners to specified torque.
- Always use new cotter pins; never reuse any cotter pin.
- Always lubricate where needed, but never overlubricate.

Steering Linkage Service

Inspect all steering linkage for bent, damaged, or broken parts, including torn or

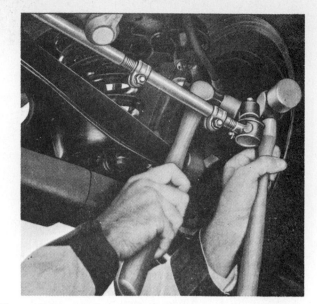

Figure 6-34. Another method of removing tie rod end stud from steering arm. *(Courtesy of Chrysler Corporation)*

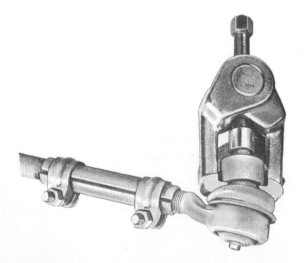

Figure 6-33. Using a special tool to remove tie rod end from steering arm after cotter pin and nut have been removed. After stud removal and after loosening clamp on adjusting sleeve, count the number of turns required to remove the tie rod end from the sleeve. Install the new tie rod end the same number of turns, and tighten clamp away from split in sleeve. *(Courtesy of Chrysler Corporation)*

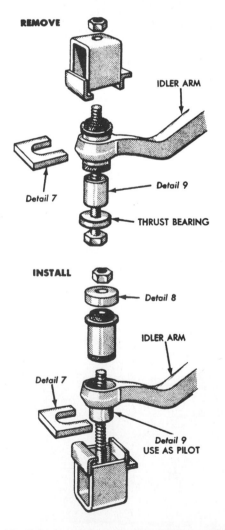

Figure 6-35. Removal (top) and installation (bottom) of one type of idler arm bushing using a special tool for the job. *(Courtesy of Chrysler Corporation)*

ruptured seals and mounting bolts and nuts. If linkage appears to be in good condition, a further inspection must be made to determine wear. Tie rod ends and idler arm bushing wear should be checked and corrected as follows:

• Support the vehicle safely on jack stands.

• Grasp the tire firmly at front and back and move back and forth alternately a short distance. Do this vigorously.

• Observe whether there is any lateral movement in each tie rod end; if so, replace the tie rod end.

• To check idler arm wear, grasp the linkage at the idler arm and try to move it up and down. If there is any movement up or down, replace the idler arm.

• Replace any bent or damaged parts.

• Tighten all bolts and nuts to specified torque and lock with new cotter pin. If holes do not line up, tighten to the next hole only; never loosen to line up cotter pin holes.

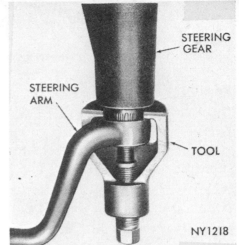

Figure 6-36. Using a puller to remove the pitman arm from the steering gear sector shaft after nut has been removed.

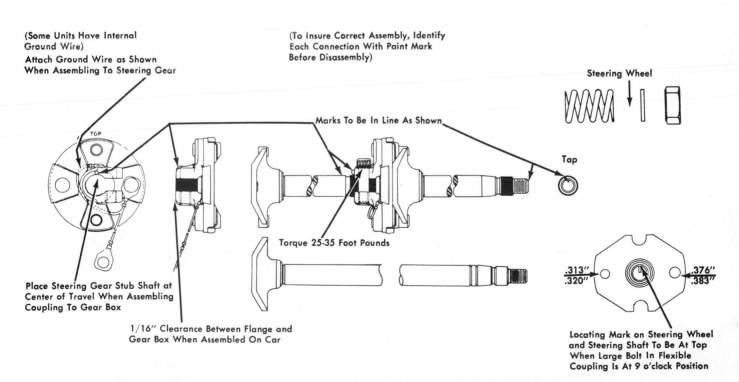

Figure 6-37. Special precautions must be observed before disassembly and during assembly to ensure correct assembly of steering column shaft and steering gear shaft connections. This is only one example; other makes and models require other procedures. *(Courtesy of General Motors Corporation)*

Manual-Steering Gear Service

The steering gear should be checked to determine if adjustment or repairs are needed. In general, if steering gear is suspected, proceed as follows.

• Check the lubricant level; if low, check for leaks and correct.

• Check mounting bolts; tighten to specifications or replace if damaged.

• Check whether there is any lost motion at the flexible coupling; correct as necessary.

• Check whether the pitman arm and nut are tight on sector shaft; tighten to specifications. If splines are worn, replace both pitman arm and sector shaft.

• Check whether there is any looseness (lost motion), roughness, or bind in the steering gear. If too loose or too tight, adjust to specifications. If adjustment does not correct the problem or if the gear is rough, overhaul the steering gear according to Figures 6-37 to 6-51 and to the manufacturer's specifications.

Figure 6-39. Removing sector shaft after side cover bolts have been removed. This is the first step in manual-steering gear disassembly. *(Courtesy of General Motors Corporation)*

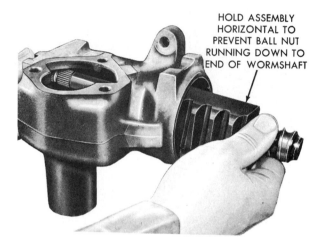

Figure 6-40. Removing wormshaft and ball-nut assembly after wormshaft bearing adjuster has been removed. *(Courtesy of General Motors Corporation)*

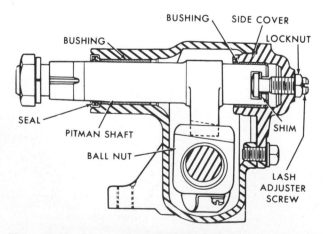

Figure 6-38. Cross-sectional view of manual steering gear. Note that sector shaft teeth and ball nut teeth are tapered. Note that the lash adjuster screw, when turned, moves the sector shaft endwise to adjust mesh. *(Courtesy of General Motors Corporation)*

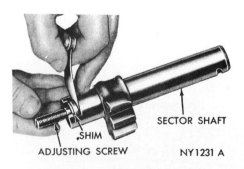

Figure 6-41. Checking lash adjusting screw head-to-sector shaft clearance with a feeler gauge. Excessive clearance at this point allows excessive sector shaft movement, which results in loose steering and vehicle wander. Use a thicker shim to restore correct clearance. *(Courtesy of Chrysler Corporation)*

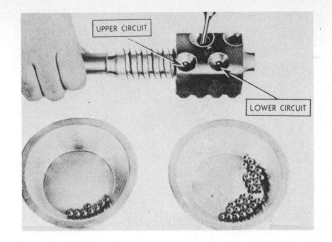

Figure 6-42. Installing balls into ball-nut assembly. Ball guides of this type (with holes for loading) make the job easy. A retainer covers the holes and keeps the ball guides in place after loading. Since there are two ball circuits, each circuit is loaded separately. Turning the wormshaft helps to load the circuits. *(Courtesy of General Motors Corporation)*

Figure 6-43. Assembling ball-nut assembly on unit that has no ball loading holes in ball guides. On this type of unit the wormshaft must not be turned during this procedure, since this will allow balls to escape from their circuit. *(Courtesy of General Motors Corporation)*

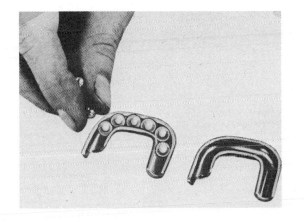

Figure 6-45. Installing loaded ball guide in previously loaded ball-nut assembly. A retainer plate holds both guides in place after installation. *(Courtesy of General Motors Corporation)*

Figure 6-44. After loading ball nut with balls (Figure 6-43), ball guides must be loaded with balls as shown here. Use some steering gear lubricant to help keep balls in place during guide installation. *(Courtesy of General Motors Corporation)*

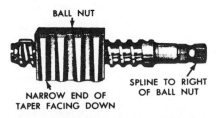

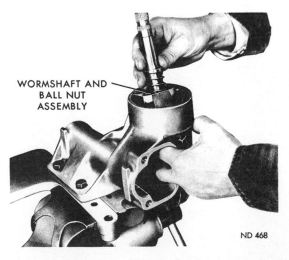

Figure 6-46. Assembled ball nut and wormshaft should look like this. If wormshaft is installed in other end of nut, proper mesh and lash adjustment cannot be achieved. *(Courtesy of Ford Motor Co. of Canada Ltd.)*

Figure 6-47. During assembly, install wormshaft and ball nut in gear case first, as shown here. *(Courtesy of Chrysler Corporation)*

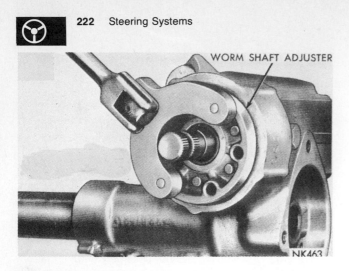

Figure 6-48. After wormshaft and ball nut are installed, adjust wormshaft bearing preload to manufacturer's specifications. *(Courtesy of Chrysler Corporation)*

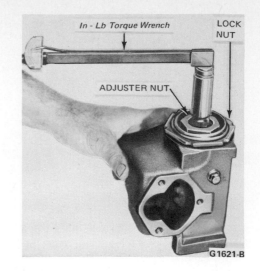

Figure 6-49. Wormshaft bearing preload is checked with a torque wrench as shown here. Too much preload causes early bearing failure, while loose bearings cause loose steering as well as early failure. *(Courtesy of Ford Motor Co. of Canada Ltd.)*

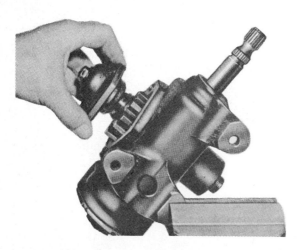

Figure 6-50. After correct worm bearing preload has been obtained, install sector shaft assembly. Make sure that lash adjuster is backed off to prevent teeth from binding when bolting down the cover. *(Courtesy of General Motors Corporation)*

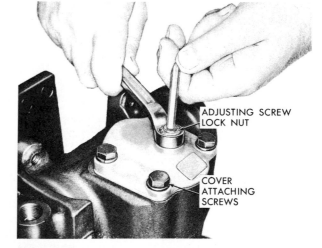

Figure 6-51. Proper lash adjustment is achieved by loosening the lock nut and turning the adjusting screw. Wormshaft should be turned back and forth through midpoint of travel to obtain "feel" of "high point" and correct lash adjustment. Follow shop manual specifications for correct setting. *(Courtesy of General Motors Corporation)*

Power Steering Service

Power steering problems are usually evidenced by one of the following conditions:

- Hard to steer
- Erratic assist
- Noisy steering
- Pulling to one side (self-steering)

To determine the cause of problems in power steering, proceed as follows.

- Check the lubricant level. If low, check for leaks and correct.
- Check belt tension and condition; replace any glazed or damaged belt.

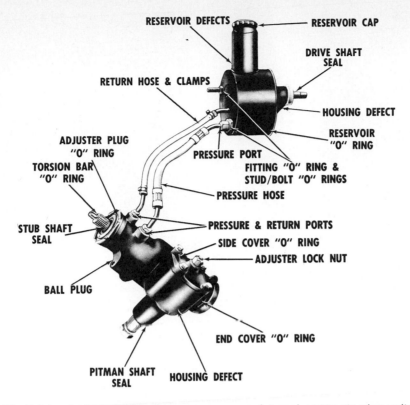

Figure 6-52. Points at which fluid leakage can occur on integral power steering units. Wipe area of suspected leakage clean, then check source of leakage. Fluid can run down onto adjacent parts and lead to incorrect diagnosis. *(Courtesy of General Motors Corporation)*

• Check pump and gear mounting bolts; correct as necessary.

• Check for lost motion at flexible coupling; correct as necessary.

• Check the pitman arm and nut for any looseness; if the pitman arm or sector shaft splines are worn, replace.

• Check for any looseness (lost motion), roughness, or bind in the steering gear; if too loose or too tight, adjust to specifications. If adjustment does not correct the problem or if gear is rough, overhaul according to the manufacturer's procedures and specifications.

• Jack the front of the car to raise front wheels off the floor, support the car on stands, position the steering wheel at the center of travel, start the engine, and, without touching the steering wheel, observe whether the wheel self-steers to right or left; if it does, adjust or repair as required by the manufacturer's manual.

• Check system pressure to determine if pressure developed meets specifications. If too low, pump must be serviced. If pressure is correct and all other checks have been completed, steering gear should be suspected. Remove and repair as needed.

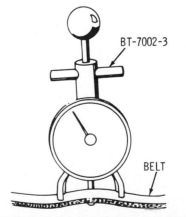

Figure 6-53. Checking belt tension on power steering with tension gauge. Incorrect tension can result in erratic steering assist, belt squeal, and early belt, bearing, and pulley failure. *(Courtesy of General Motors Corporation)*

Figure 6-54. Power-steering fluid level should be checked with system at operating temperature and according to the manufacturer's specifications.

Figure 6-55. Checking power-steering system pressure is one method of problem diagnosis. This will determine if pressure is up to the manufacturer's specifications and whether the pump or steering gear unit is at fault.

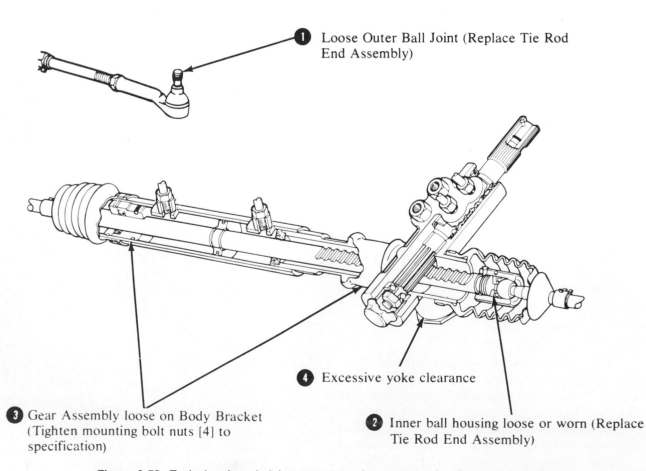

1 Loose Outer Ball Joint (Replace Tie Rod End Assembly)

4 Excessive yoke clearance

3 Gear Assembly loose on Body Bracket (Tighten mounting bolt nuts [4] to specification)

2 Inner ball housing loose or worn (Replace Tie Rod End Assembly)

Figure 6-56. Typical rack-and-pinion power-steering gear wander diagnosis procedure.

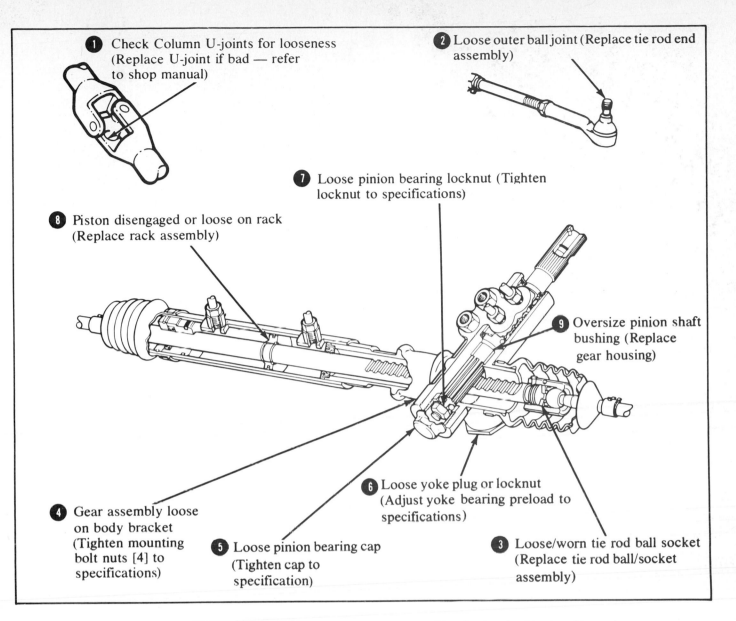

1 Check Column U-joints for looseness (Replace U-joint if bad — refer to shop manual)

2 Loose outer ball joint (Replace tie rod end assembly)

7 Loose pinion bearing locknut (Tighten locknut to specifications)

8 Piston disengaged or loose on rack (Replace rack assembly)

9 Oversize pinion shaft bushing (Replace gear housing)

6 Loose yoke plug or locknut (Adjust yoke bearing preload to specifications)

4 Gear assembly loose on body bracket (Tighten mounting bolt nuts [4] to specifications)

5 Loose pinion bearing cap (Tighten cap to specification)

3 Loose/worn tie rod ball socket (Replace tie rod ball/socket assembly)

Figure 6-57. Diagnosis procedure for steering feedback condition (rattle, chuckle, knocking noises in steering gear) on power rack-and-pinion steering gear. *(Courtesy of Ford Motor Co. of Canada Ltd.)*

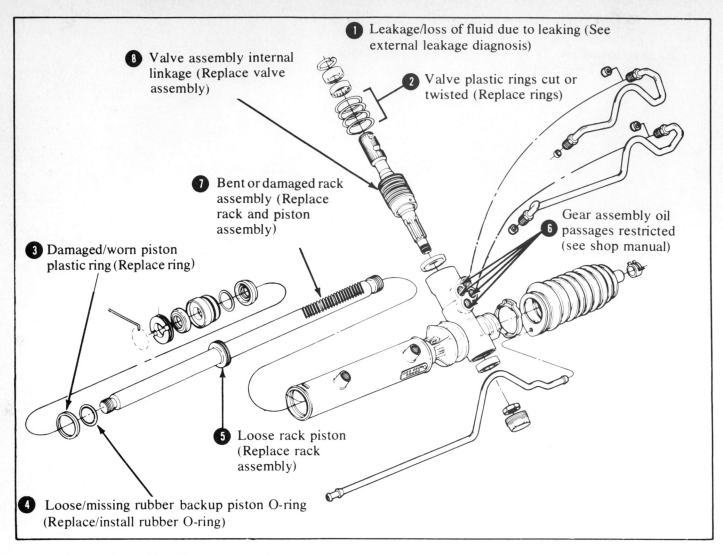

1 Leakage/loss of fluid due to leaking (See external leakage diagnosis)

8 Valve assembly internal linkage (Replace valve assembly)

2 Valve plastic rings cut or twisted (Replace rings)

7 Bent or damaged rack assembly (Replace rack and piston assembly)

6 Gear assembly oil passages restricted (see shop manual)

3 Damaged/worn piston plastic ring (Replace ring)

5 Loose rack piston (Replace rack assembly)

4 Loose/missing rubber backup piston O-ring (Replace/install rubber O-ring)

Figure 6-58. Diagnosis procedure for heavy steering effort condition (poor assist or loss of assist) for power rack-and-pinion steering gear. *(Courtesy of Ford Motor Co. of Canada Ltd.)*

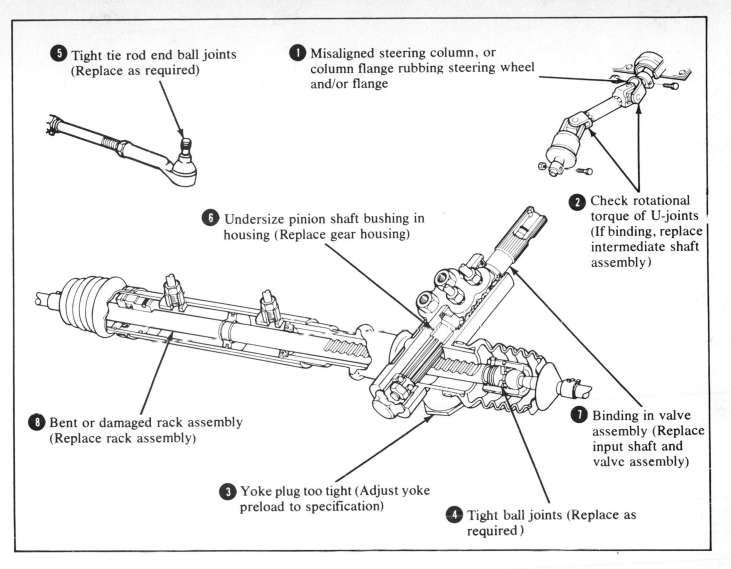

5 Tight tie rod end ball joints (Replace as required)

1 Misaligned steering column, or column flange rubbing steering wheel and/or flange

2 Check rotational torque of U-joints (If binding, replace intermediate shaft assembly)

6 Undersize pinion shaft bushing in housing (Replace gear housing)

8 Bent or damaged rack assembly (Replace rack assembly)

7 Binding in valve assembly (Replace input shaft and valve assembly)

3 Yoke plug too tight (Adjust yoke preload to specification)

4 Tight ball joints (Replace as required)

Figure 6-59. Diagnosis procedure for poor returnability condition (stick feeling) for power rack-and-pinion steering gear. *(Courtesy of Ford Motor Co. of Canada Ltd.)*

STEERING SYSTEMS

**TITLE: <u>MANUAL STEERING
GEARS AND LINKAGE</u>**

STUDENT'S NAME _____

PERFORMANCE OBJECTIVES

After sufficient opportunity to study this portion of the text and the appropriate training models and with the instructor's supervision and demonstrations, you should be able to perform the following tasks at the request of your instructor.

TASK 1---State the purpose of the steering linkage. Describe the construction and operation of the parallelogram type and the rack-and-pinion type of steering linkage under normal and abnormal operating conditions. Name the components indicated on the steering linkage diagrams.

TASK 2---State the purpose of the manual steering gear. Describe the construction and operation of the manual steering gear under normal and abnormal operating conditions. Name the components indicated on the manual-steering gear diagrams.

PERFORMANCE EVALUATION

Your instructor may require you to perform these tasks in any of the following ways in order to evaluate your performance:
- By asking test questions
- By asking you to describe the performance of these tasks in writing
- By asking you to describe the performance of these tasks orally

Chapter 6

PROJECT NO. 6-B

TITLE: <u>POWER-STEERING</u>
<u>SYSTEM</u>

STUDENT'S NAME _____

PERFORMANCE OBJECTIVES

After sufficient opportunity to study this portion of the text and the appropriate training models and with the instructor's supervision and demonstrations, you should be able to perform the following tasks at the request of your instructor.

TASK 1---State the purpose of the power-steering system. Describe the basic construction and operation of the integral, the linkage booster, and the rack-and-pinion types of power-steering systems.

TASK 2---Name the components indicated on the power-steering gear and pump diagrams.

PERFORMANCE EVALUATION

Your instructor may require you to perform these tasks in any of the following ways in order to evaluate your performance:
- By asking test questions
- By asking you to describe the performance of these tasks in writing
- By asking you to describe the performance of these tasks orally

TITLE: <u>STEERING SYSTEM</u>
<u>SERVICE</u>

STUDENT'S NAME _____

PERFORMANCE OBJECTIVES

After sufficient opportunity to study this portion of the text and the appropriate training models and with the instructor's supervision and demonstrations, you should be able to perform the following tasks at the request of your instructor.

TASK 1---Complete the diagnostic chart provided. Diagnose steering system problems on a vehicle specified by your instructor.

TASK 2---Repair the steering linkage on a vehicle designated by your instructor. Follow the procedures and specifications given in the text, in the shop manual, and by your instructor. Use the worksheet provided to record the results.

TASK 3---Remove, disassemble, clean, inspect, repair, assemble, adjust, and install a manual-steering gear specified by your instructor. Use the worksheet provided to record the results.

TASK 4---Remove, disassemble, clean, inspect, repair, assemble, and install a power-steering pump designated by your instructor. Use the worksheet provided to record the results.

TASK 5---Remove, disassemble, clean, inspect, repair, assemble, adjust, and install a rack-and-pinion power-steering unit specified by your instructor. Use the worksheet provided to record the results.

TASK 6---Remove, disassemble, clean, inspect, repair, assemble, adjust, and install an integral power-steering gear specified by your instructor. Use the worksheet provided to record the results.

PERFORMANCE EVALUATION

Your ability to perform these tasks will be evaluated by your instructor on the basis of how accurately you have followed the procedures and specifications given to you by your instructor, by the text, and by the manufacturer's service manual.

PROBLEM	CAUSE	CORRECTION
Excessive play in steering	1.	1.
	2.	2.
	3.	3.
	4.	4.
	5.	5.
	6.	6.
Hard Steering	1.	1.
	2.	2.
	3.	3.
	4.	4.
	5.	5.
	6.	6.
Car pulls to one side	1.	1.
Front-wheel shimmy	1.	1.

— PITMAN ARM   OK _____   NOT OK _____   EXPLAIN _____

— CENTER LINK
  (RELAY ROD)   OK _____   NOT OK _____   EXPLAIN _____

— IDLER ARM   OK _____   NOT OK _____   EXPLAIN _____

— TIE ROD ENDS   OK _____   NOT OK _____   EXPLAIN _____
  LEFT OUTER   OK _____   NOT OK _____   EXPLAIN _____
  LEFT INNER   OK _____   NOT OK _____   EXPLAIN _____
  RIGHT OUTER   OK _____   NOT OK _____   EXPLAIN _____
  RIGHT INNER   OK _____   NOT OK _____   EXPLAIN _____

— ADJUSTING SLEEVES   OK _____   NOT OK _____   EXPLAIN _____

— SLEEVE CLAMPS   OK _____   NOT OK _____   EXPLAIN _____

— STEERING ARMS   OK _____   NOT OK _____   EXPLAIN _____

STUDENT SIGN HERE   _____

INSTRUCTOR VERIFY   _____

BEFORE REMOVAL
— LEAKAGE   OK _____   NOT OK _____
— LUBRICANT LEVEL   OK _____   NOT OK _____
— MOUNTING   OK _____   NOT OK _____
— SECTOR SHAFT AND
   BEARINGS   OK _____   NOT OK _____
— ADJUSTMENT   OK _____   NOT OK _____

AFTER REMOVAL AND DISASSEMBLY
— COVER PLATE ASSEMBLY OK _____   NOT OK _____
— HOUSING   OK _____   NOT OK _____
— SECTOR SHAFT   OK _____   NOT OK _____
   EXPLAIN _____
— WORM SHAFT   OK _____   NOT OK _____
   EXPLAIN _____
— WORM SHAFT BEARINGS   OK _____   NOT OK _____
   EXPLAIN _____
— BALL NUT ASSEMBLY   OK _____   NOT OK _____
   EXPLAIN _____

ASSEMBLY
— WORM BEARING PRELOAD OK _____   NOT OK _____
— LASH ADJUSTMENT   OK _____   NOT OK _____
— LUBRICANT LEVEL   OK _____   NOT OK _____
— FINAL INSPECTION   OK _____   NOT OK _____
— INSTALLATION   OK _____   NOT OK _____
— OPERATION   OK _____   NOT OK _____

STUDENT SIGN HERE _____

INSTRUCTOR VERIFY _____

BEFORE REMOVAL

— LEAKAGE   OK _____   NOT OK _____   EXPLAIN _____

— BELLOWS SEALS   OK _____   NOT OK _____   EXPLAIN _____

— MOUNTING   OK _____   NOT OK _____   EXPLAIN _____

— ADJUSTMENT   OK _____   NOT OK _____   EXPLAIN _____

AFTER REMOVAL AND DISASSEMBLY

— HOUSING   OK _____   NOT OK _____   EXPLAIN _____

— RACK   OK _____   NOT OK _____   EXPLAIN _____

— RACK SLIPPER AND

    SPRING   OK _____   NOT OK _____   EXPLAIN _____

— PINION   OK _____   NOT OK _____   EXPLAIN _____

— PINION BEARINGS   OK _____   NOT OK _____   EXPLAIN _____

— COVER PLATE AND SHIMS OK _____   NOT OK _____   EXPLAIN _____

ASSEMBLY

— LUBRICATION   OK _____   NOT OK _____   EXPLAIN _____

— ADJUSTMENT   OK _____   NOT OK _____   EXPLAIN _____

— INSTALLATION   OK _____   NOT OK _____   EXPLAIN _____

STUDENT SIGN HERE   _____

INSTRUCTOR VERIFY   _____

BEFORE REMOVAL

— LEAKAGE   OK _____   NOT·OK _____   EXPLAIN _____

— PULLEY   OK _____   NOT OK _____   EXPLAIN _____

— MOUNTING   OK _____   NOT OK _____   EXPLAIN _____

— FLUID LEVEL   OK _____   NOT OK _____   EXPLAIN _____

— BELT   OK _____   NOT OK _____   EXPLAIN _____

AFTER REMOVAL AND DISASSEMBLY

— RESERVOIR   OK _____   NOT OK _____   EXPLAIN _____

— PUMP BODY   OK _____   NOT OK _____   EXPLAIN _____

— SHAFT   OK _____   NOT OK _____   EXPLAIN _____

— BEARINGS   OK _____   NOT OK _____   EXPLAIN _____

— ROLLERS , ROTORS,
SLIPPERS   OK _____   NOT OK _____   EXPLAIN _____

— PRESSURE PLATES   OK _____   NOT·OK _____   EXPLAIN _____

— CAM RING   OK _____   NOT OK _____   EXPLAIN _____

— CONTROL VALVE
ASSEMBLY   OK _____   NOT OK _____   EXPLAIN _____

— ASSEMBLY   OK _____   NOT OK _____   EXPLAIN _____

— INSTALLATION   OK _____   NOT OK _____   EXPLAIN _____

— OPERATION   OK _____   NOT OK _____   EXPLAIN _____

STUDENT SIGN HERE _____

INSTRUCTOR VERIFY _____

BEFORE REMOVAL
— LEAKAGE   OK _____   NOT OK _____   EXPLAIN _____

— BELLOWS SEALS   OK _____   NOT OK _____   EXPLAIN _____

— MOUNTING   OK _____   NOT OK _____   EXPLAIN _____

— ADJUSTMENT   OK _____   NOT OK _____   EXPLAIN _____

AFTER REMOVAL AND DISASSEMBLY
— LINES AND FITTINGS   OK _____   NOT OK _____   EXPLAIN _____

— HOUSING   OK _____   NOT OK _____   EXPLAIN _____

— PISTONS AND SEALS   OK _____   NOT OK _____   EXPLAIN _____

— RACK   OK _____   NOT OK _____   EXPLAIN _____

— RACK SLIPPERS AND
SPRING   OK _____   NOT OK _____   EXPLAIN _____

— PINION   OK _____   NOT OK _____   EXPLAIN _____

— PINION BEARINGS   OK _____   NOT OK _____   EXPLAIN _____

— COVER PLATE AND
SHIMS   OK _____   NOT OK _____   EXPLAIN _____

— VALVE ASSEMBLY   OK _____   NOT OK _____   EXPLAIN _____

ASSEMBLY
— LUBRICATION   OK _____   NOT OK _____   EXPLAIN _____

— INSTALLATION   OK _____   NOT OK _____   EXPLAIN _____

— ADJUSTMENT   OK _____   NOT OK _____   EXPLAIN _____

STUDENT SIGN HERE _____

INSTRUCTOR VERIFY _____

BEFORE REMOVAL
— LEAKAGE   OK _____   NOT OK _____   EXPLAIN _____
— MOUNTING   OK _____   NOT OK _____   EXPLAIN _____
— LINES AND FITTINGS   OK _____   NOT OK _____   EXPLAIN _____

AFTER REMOVAL AND DISASSEMBLY
— COVER PLATE ASSEMBLY OK _____   NOT OK _____   EXPLAIN _____
— HOUSING   OK _____   NOT OK _____   EXPLAIN _____
— SECTOR SHAFT   OK _____   NOT OK _____   EXPLAIN _____
— SECTOR SHAFT BEARINGS OK _____   NOT OK _____   EXPLAIN _____
— WORM SHAFT   OK _____   NOT OK _____   EXPLAIN _____
— WORM SHAFT BEARINGS   OK _____   NOT OK _____   EXPLAIN _____
— RACK-PISTON ASSEMBLY   OK _____   NOT OK _____   EXPLAIN _____
— CONTROL VALVE AND
   REACTION ASSEMBLY   OK _____   NOT OK _____   EXPLAIN _____

ASSEMBLY
— WORM BEARING PRELOAD OK _____   NOT OK _____   EXPLAIN _____
— LASH ADJUSTMENT   OK _____   NOT OK _____   EXPLAIN _____
— VALVE ADJUSTMENT   OK _____   NOT OK _____   EXPLAIN _____
— FINAL INSPECTION
   BEFORE INSTALLATION   OK _____   NOT OK _____   EXPLAIN _____
— INSTALLATION   OK _____   NOT OK _____   EXPLAIN _____
— FLUID LEVEL   OK _____   NOT OK _____   EXPLAIN _____
— BELT TENSION   OK _____   NOT OK _____   EXPLAIN _____
— LINES AND FITTINGS,
   ROUTING AND
   SUPPORTS   OK _____   NOT OK _____   EXPLAIN _____
— OPERATION   OK _____   NOT OK _____   EXPLAIN _____

STUDENT SIGN HERE _____

INSTRUCTOR VERIFY _____

Self-Check

1. What part of the steering linkage provides for linkage adjustment?
2. What is the purpose of the idler arm?
3. The outer tie rod ends are connected to the_____.
4. What is the purpose of using the recirculating ball-and-nut design in the manual steering gear?
5. The rack-and-pinion steering gear is an integral part of the_____.
6. List three types of power-steering pumps.
7. List three types of power-steering systems.
8. What is the purpose of the flexible coupling between the steering gear and steering column shaft?
9. List three causes of hard steering.
10. List four causes of excessive play in steering.

Performance Evaluation

After you have thoroughly studied this chapter and had sufficient practice on steering system components, you should be able to do the following.

1. Follow the accepted general precautions.

2. Correctly disassemble all steering system components.

3. Properly clean all steering system components.

4. Accurately inspect and measure all steering system components to determine their serviceability.

5. Properly replace all steering system components according to the manufacturer's specifications.

6. Perform the necessary inspection to determine the success of the steering system and component overhaul.

7. Properly prepare the vehicle for customer acceptance.

8. Complete the Self-Check with at least 80 percent accuracy.

9. Complete all practical work with 100 percent accuracy.

Chapter 7

Wheel Alignment and Balance

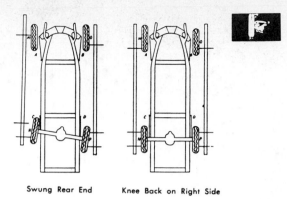

Swung Rear End   Knee Back on Right Side

Figure 7-1. Rear wheels must follow front wheels properly for good steering, handling, and vehicle stability. This is called *tracking*, which occurs if all four wheels are parallel to the frame. This illustration shows the method for checking tracking by using a track bar (sometimes called *tramming*). When track bar pointers are set to contact wheels on one side of vehicle, they should contact wheels at same points on the other side. If they do not, there is possibly a shifted frame or suspension parts.

The purpose of proper wheel alignment and balance is to provide maximum safety, ease of handling, stability, and directional control of the vehicle. This requires that each of the steering angles (steering geometry) be adjusted to the specifications recommended by the vehicle manufacturer. Different makes and models of vehicles require different settings. Follow the specific shop manual for each vehicle. The wheels must also be in proper dynamic and static balance to achieve these purposes.

PART 1 STEERING GEOMETRY

Tracking

For proper tracking, all four wheels must be parallel to the frame. This requires that the wheelbase be equal on both sides of the vehicle. The four wheels should be positioned to form a rectangle.

Camber

Camber is the inward or outward tilt of the wheel at the top. Inward tilt is negative camber and outward tilt is positive camber.

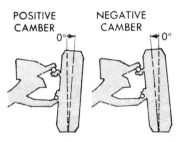

Figure 7-2. Camber is the inward or outward tilt of the wheel at the top. If the wheel is absolutely vertical, there is zero camber. *(Courtesy of Chrysler Corporation)*

WHEEL WITH NEGATIVE CAMBER TENDS TO ROLL UPHILL

Figure 7-3. Effect of negative camber on directional control. If there is a difference in camber from one front wheel to the other, the vehicle will tend to pull to the side with the most positive camber. *(Courtesy of Bear)*

The tilt of the wheel (camber) is measured in degrees and is adjustable on many vehicles. Camber is needed for the following purposes:

1. To bring the point of load more nearly to the center of the tire where it contacts the road

2. To reduce steering effort by putting more of the load at the inner end of the spindle on the larger bearing

3. To reduce tire wear

Incorrect camber settings can cause the following:

1. Excessive wear to suspension parts

2. Excessive wheel bearing wear

3. Excessive tire wear (excess positive camber, outside tire-tread wear; excess negative camber, inside tire-tread wear)

4. Excessive unequal camber will cause the car to pull to one side

5. Excessive negative camber on the right wheel will cause pull to the left

6. Excessive positive camber on the right wheel will cause pull to the right

7. Excessive negative camber on the left wheel will cause pull to the right

8. Excessive positive camber on the left wheel will cause pull to the left

Caster

Caster is the forward or backward tilt of the spindle or steering knuckle at the top when viewed from the side. Forward tilt is negative caster and backward tilt is positive caster. Caster is measured in the number of degrees that it is forward or backward from true vertical and is adjustable on many vehicles. Caster is needed for the following purposes:

1. To aid in directional control by helping to keep wheels in straight-ahead position on some models

2. To help return the wheels to straight ahead after a turn

3. To offset the effects of road crown by setting the left side 1/4° to 1/2° more negative (or less positive)

Incorrect caster can result in the following:

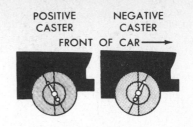

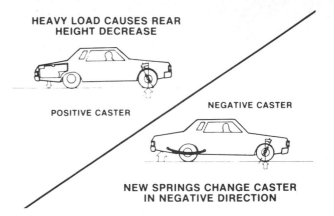

Figure 7-4. Caster is the forward or backward tilt of the spindle (steering knuckle) at the top as viewed from the side. When the spindle support is vertical, there is zero caster. *(Courtesy of Chrysler Corporation)*

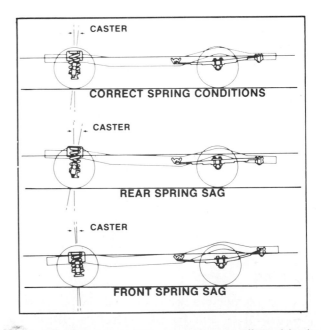

Figure 7-5. Caster is affected by vehicle load and by sagged springs. *(Courtesy of Ford Motor Co. of Canada Ltd.)*

Figure 7-6. Steering angles are adversely affected by incorrect suspension height. Sagged springs are a common cause of this condition. *(Courtesy of Ford Motor Co. of Canada Ltd.)*

1. Wander (too little caster)

2. Hard steering (too much caster)

3. Road shock and shimmy (too much caster)

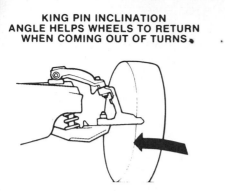

KING PIN INCLINATION
ANGLE HELPS WHEELS TO RETURN
WHEN COMING OUT OF TURNS.

Figure 7-10. Another effect of steering axis inclination is shown here. *(Courtesy of Chrysler Corporation)*

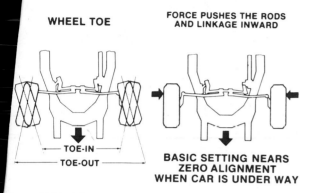

WHEEL TOE

FORCE PUSHES THE RODS
AND LINKAGE INWARD

TOE-IN
TOE-OUT

BASIC SETTING NEARS
ZERO ALIGNMENT
WHEN CAR IS UNDER WAY

Figure 7-11. The objective in setting toe is to have the front wheels running straight ahead (zero running toe). To achieve this, a small amount of static toe-in must be present. Driving force deflects steering parts enough to result in zero running toe. Some cars with a negative scrub radius may require a toe-out setting. *(Courtesy of Ford Motor Co. of Canada Ltd.)*

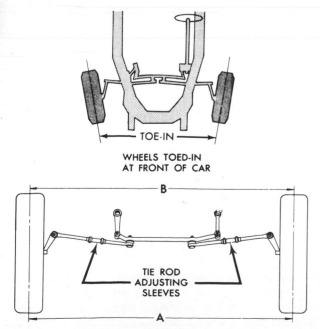

TOE-IN

WHEELS TOED-IN
AT FRONT OF CAR

B

TIE ROD
ADJUSTING
SLEEVES

A

Figure 7-12. When front wheels have toe-in, they are closer together at the front wheels than at the rear. Dimension A is less than dimension B. *(Courtesy of General Motors Corporation and Chrysler Corporation)*

together slightly at the front when driving down the road, depending on vehicle design. This provides a zero running toe and no tire scuffing.

Incorrect toe-in or toe-out is the most frequent cause of rapid tire-tread wear. Toe setting is the last adjustment to be made when performing a wheel alignment. On most front-wheel-drive vehicles, toe-out setting is required to provide a zero running toe. This is because the driving front wheels are trying to go around the SAI pivot point with a negative scrub radius.

Toe-Out on Turns

Toe-out on turns is the different turning radius of the two front wheels. When the car is in a turn, the inner wheel is turned more than the outer wheel, resulting in toe-out on turns. This is caused by the steering arms being bent inward where they connect to the steering linkage. It is needed to prevent tire scuffing (dragging sideways) during a turn. Since the inner wheel follows a smaller circle than the outer wheel when in a turn, toe-out on turns is necessary. Toe-out on turns is not adjustable and is corrected by replacing steering arms.

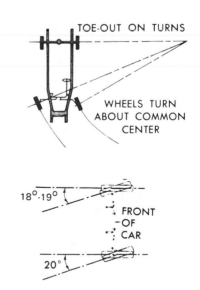

TOE-OUT ON TURNS

WHEELS TURN
ABOUT COMMON
CENTER

18°-19°

FRONT
OF
CAR

20°

Figure 7-13. Toe-out on turns (turning radius) is needed to reduce tire scuffing when turning corner. Since all four wheels turn around a common center, the inner front wheel must be turned sharper, as shown above. This is accomplished by steering arm design. Steering arms are angled inward where they attach to the steering linkage. *(Courtesy of General Motors Corporation and Chrysler Corporation)*

Steering Axis Inclination

Steering axis inclination is the inward tilt of the steering knuckle at the top. Steering axis inclination is measured in degrees and is not adjustable. If incorrect, suspension parts are at fault and must be replaced. Steering axis inclination is needed for the following purposes:

1. To place the weight of the vehicle more nearly under the center of the tire where it contacts the road

2. To reduce steering effort, because of item 1

3. To provide good directional control (Figure 7-10)

Incorrect steering axis inclination will affect all the preceding and may prevent being able to set camber properly.

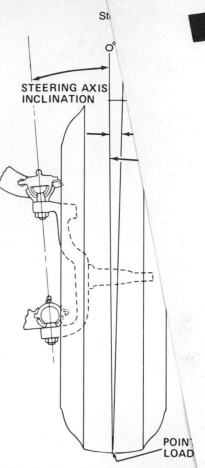

Figure 7-8. Relationship of steering axis in ber. One of the effects of these two angles point of vehicle load at the center of the t surface. *(Courtesy of General Motors Corpo*

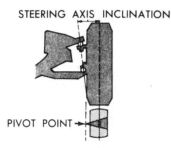

Figure 7-7. Steering axis inclination is the inward tilt of the spindle support at the top. *(Courtesy of Chrysler Corporation)*

Toe-In

Toe-in occurs when the front w slightly closer together at the front the rear. Toe-in is measured in incl limeters, or degrees. A limited am toe-in or toe-out is needed to allow fact that the wheels spread apart o

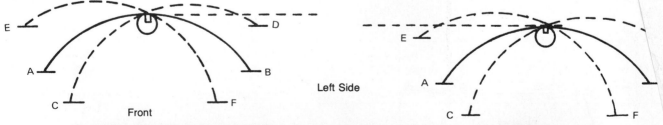

Figure 7-9. The effect of steering axis inclination on the travel of the outer end of spindle as shown by the line A to B when front wheels are turned from right to left. Since the spindle cannot in fact push the wheel into the road surface, the result is that the car is raised when turned to right or left. This helps return front wheels to straight head after a turn and aids directional stability. The line A to B represents a zero caster setting, the line C to D a negative caster setting, and the line E to F a positive caster setting. A car will lead to the side with the most negative or least positive caster. The figure on the left represents a zero camber setting; the figure on the right shows the result of a positive camber setting. The effect is that it tries to lower the outer end of the spindle.

DIAGNOSTIC CHART

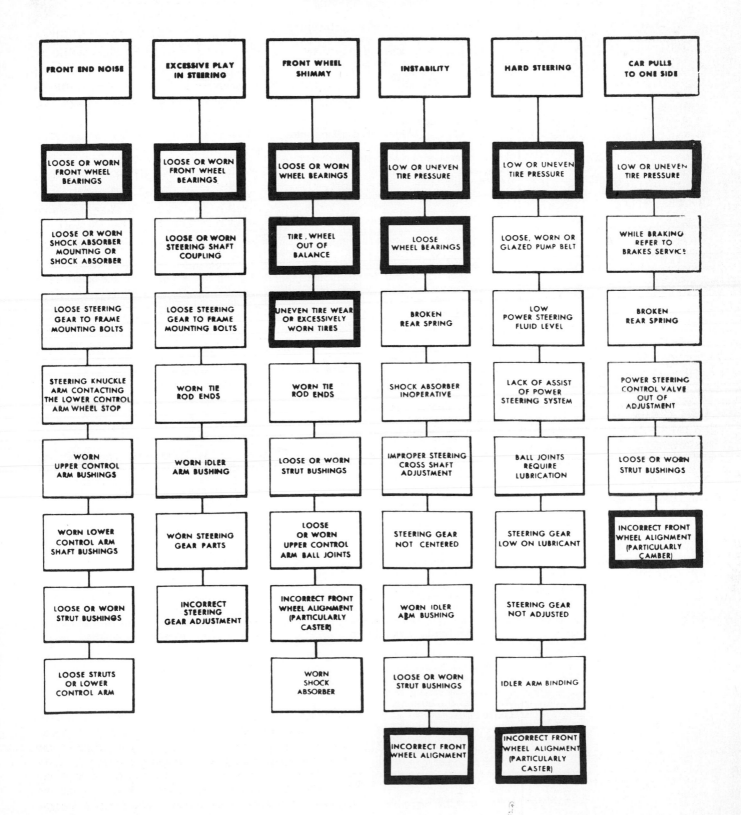

FRONT END NOISE	EXCESSIVE PLAY IN STEERING	FRONT WHEEL SHIMMY	INSTABILITY	HARD STEERING	CAR PULLS TO ONE SIDE
LOOSE OR WORN FRONT WHEEL BEARINGS	LOOSE OR WORN FRONT WHEEL BEARINGS	LOOSE OR WORN WHEEL BEARINGS	LOW OR UNEVEN TIRE PRESSURE	LOW OR UNEVEN TIRE PRESSURE	LOW OR UNEVEN TIRE PRESSURE
LOOSE OR WORN SHOCK ABSORBER MOUNTING OR SHOCK ABSORBER	LOOSE OR WORN STEERING SHAFT COUPLING	TIRE, WHEEL OUT OF BALANCE	LOOSE WHEEL BEARINGS	LOOSE, WORN OR GLAZED PUMP BELT	WHILE BRAKING REFER TO BRAKES SERVICE
LOOSE STEERING GEAR TO FRAME MOUNTING BOLTS	LOOSE STEERING GEAR TO FRAME MOUNTING BOLTS	UNEVEN TIRE WEAR OR EXCESSIVELY WORN TIRES	BROKEN REAR SPRING	LOW POWER STEERING FLUID LEVEL	BROKEN REAR SPRING
STEERING KNUCKLE ARM CONTACTING THE LOWER CONTROL ARM WHEEL STOP	WORN TIE ROD ENDS	WORN TIE ROD ENDS	SHOCK ABSORBER INOPERATIVE	LACK OF ASSIST OF POWER STEERING SYSTEM	POWER STEERING CONTROL VALVE OUT OF ADJUSTMENT
WORN UPPER CONTROL ARM BUSHINGS	WORN IDLER ARM BUSHING	LOOSE OR WORN STRUT BUSHINGS	IMPROPER STEERING CROSS SHAFT ADJUSTMENT	BALL JOINTS REQUIRE LUBRICATION	LOOSE OR WORN STRUT BUSHINGS
WORN LOWER CONTROL ARM SHAFT BUSHINGS	WORN STEERING GEAR PARTS	LOOSE OR WORN UPPER CONTROL ARM BALL JOINTS	STEERING GEAR NOT CENTERED	STEERING GEAR LOW ON LUBRICANT	INCORRECT FRONT WHEEL ALIGNMENT (PARTICULARLY CAMBER)
LOOSE OR WORN STRUT BUSHINGS	INCORRECT STEERING GEAR ADJUSTMENT	INCORRECT FRONT WHEEL ALIGNMENT (PARTICULARLY CASTER)	WORN IDLER ARM BUSHING	STEERING GEAR NOT ADJUSTED	
LOOSE STRUTS OR LOWER CONTROL ARM		WORN SHOCK ABSORBER	LOOSE OR WORN STRUT BUSHINGS	IDLER ARM BINDING	
			INCORRECT FRONT WHEEL ALIGNMENT	INCORRECT FRONT WHEEL ALIGNMENT (PARTICULARLY CASTER)	

PROBLEM	CAUSE	CORRECTION
Hard steering.	1. Ball joints require lubrication. 2. Low or uneven tire pressure. 3. Low power-steering fluid level. 4. Lack of assist of power-steering system. *5. Incorrect front-wheel alignment (particularly caster) resulting from a bent control arm, steering knuckle, or steering knuckle arm. 6. Steering gear low on lubricant. 7. Steering gear not adjusted. 8. Idler arm binding.	1. Lubricate ball joints. 2. Inflate tires to recommended pressures. 3. Fill pump reservoir to correct level. 4. Inspect, test, and service the power-steering pump and gear as required. 5. Replace bent parts and adjust the front-wheel alignment. 6. Fill gear to correct level. 7. Adjust steering gear. 8. Free up idler arm.
Car pulls to one side.	*1. Low or uneven tire pressure. 2. Front brake dragging. 3. Grease, lubricant, or brake fluid leaking onto brake lining. 4. Loose or excessively worn strut bushings. 5. Power-steering control valve out of adjustment. *6. Incorrect front-wheel alignment (particularly camber). 7. Broken or weak rear spring.	1. Inflate tires to recommended pressure. 2. Adjust brakes. 3. Replace brake shoe and lining as necessary and stop all leaks. 4. Tighten or replace strut bushings. 5. Adjust steering gear control valve. 6. Adjust front-wheel alignment. 7. Replace spring.
Excessive play in steering.	1. Worn or loose front-wheel bearings. 2. Incorrect steering gear adjustment. 3. Loose steering gear-to-frame mounting bolts. 4. Worn ball joints or tie rod. 5. Worn steering gear parts. 6. Worn upper or lower ball joints.	1. Adjust or replace wheel bearings as necessary. 2. Adjust steering gear. 3. Tighten steering gear to frame bolts. 4. Replace ball joints or tie rods as necessary. 5. Replace worn steering gear parts and adjust as necessary. 6. Replace ball joints.
Front-wheel shimmy.	*1. Tire, wheel out of balance. 2. Uneven tire wear or excessively worn tires. 3. Worn or loose wheel bearings. 4. Worn tie rod ends. 5. Strut mounting bushings loose or worn. *6. Incorrect front-wheel alignment (particularly caster). 7. Worn or loose upper control arm ball joints.	1. Balance wheel and tire assembly. 2. Rotate or replace tires as necessary. 3. Replace or adjust wheel bearings as necessary. 4. Replace tie rod ends. 5. Replace strut mounting bushings. 6. Adjust front-wheel alignment. 7. Inspect ball joints and replace where required.
Front-end noise.	1. Ball joint needs lubrication. 2. Shock absorber and bushings worn. 3. Worn strut bushings. 4. Loose struts, lower control arm bolts and nuts 5. Loose steering gear on frame. 6. Worn upper control arm bushings. 7. Worn lower control arm shaft bushings. 8. Worn upper or lower ball joint. 9. Worn tie rod ends. 10. Loose or worn front-wheel bearings. *11. Steering knuckle arm contacting the lower control arm, strut, or wheel stop.	1. Lubricate ball joint. 2. Replace bushings or shock. 3. Replace bushing. 4. Tighten all bolts and nuts. 5. Tighten the steering gear mounting bolts. 6. Replace worn bushings. 7. Replace worn bushings. 8. Replace ball joint. 9. Replace tie rod end. 10. Adjust or replace bearings as necessary. 11. Smooth off the contacting area and lubricate with a water-resistant grease.
Instability.	*1. Low or uneven tire pressure. 2. Loose wheel bearings. 3. Improper steering cross-shaft adjustment. 4. Steering gear not centered. 5. Worn idler arm bushing. 6. Loose or excessively worn front strut bushings. 7. Weak or broken rear spring. *8. Incorrect front-wheel alignment. 9. Shock absorber inoperative.	1. Inflate tires to correct pressure. 2. Adjust wheel bearing. 3. Adjust steering cross shaft. 4. Adjust steering gear. 5. Replace bushing. 6. Tighten nuts on strut rods or replace bushings. 7. Replace spring. 8. Measure and adjust front-wheel alignment. 9. Replace shock absorber.

*Wheel alignment problems. However, when diagnosing alignment problems, other related items must also be considered.

PART 3 WHEEL ALIGNMENT PROCEDURE

General Precautions

Customer and vehicle safety depend on the technician's ability to follow proper procedures and specifications. To achieve this, the following factors should be included.

• Perform all prealignment checks properly to determine extent of repairs required.

• The vehicle's steering and suspension system, including tires, should be in good condition before attempting alignment.

• Use all alignment equipment as recommended by manufacturer.

• Tighten all fasteners to specified torque.

• Install cotter pins wherever required.

• Observe all safety precautions when positioning the vehicle on the alignment machine.

Prealignment Checks

The following checks should be made before attempting wheel alignment. Correct any abnormal conditions before alignment.

1. Check tire size (tires should be the size and type recommended by the manufacturer).

2. Check tire wear (badly worn tires will affect steering and handling).

3. Correct all tire pressures.

4. Check wheels and tires for radial and lateral run-out. See Chapter 5 for the procedure.

5. The vehicle should be at curb weight (all accessories in place, full tank of fuel, spare tire in place, no passengers or additional weight).

6. Check the suspension system condition front and rear. See Chapter 4 for the procedure.

7. Check the steering linkage and gear condition. See Chapter 6 for the procedure.

8. Lubricate all lubrication points on the steering and suspension systems.

9. Balance all wheels. See Chapter 5 for the procedure.

Adjustment Methods

Adjustment methods vary considerably from one vehicle to another. Some use shims; others use eccentrics, slotted hole adjustments, or adjustable rods. Camber, caster, and toe-in are adjustable on most vehicles. Conversion kits are available for many nonadjustable strut suspension systems to allow adjustment for caster and camber correction. Steering axis inclination and toe-out on turns are not adjustable. A number of different points of adjustment are illustrated.

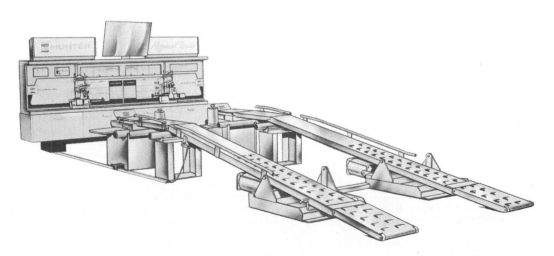

Figure 7-14. Typical wheel alignment machine mounted above the floor. *(Courtesy of Hunter Engineering Co.)*

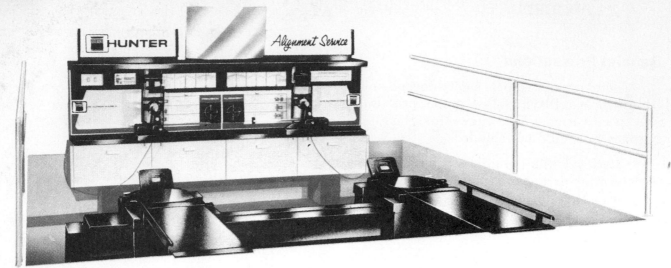

Figure 7-15. Wheel alignment machine, pit-mounted type. *(Courtesy of Hunter Engineering Co.)*

● **HERE'S TOMORROW'S ALIGN-MENT**...Electron-A-Line's advanced analytical alignment sensing system allows rapidly feeding alignment data into master computerized complete-vehicle automotive diagnostic service systems.

● **AUTOMATIC COMPENSATION**... Simply attach sensor units to wheels, and by rotating wheel one-half revolution and back to the original position, corrections for wheel assembly discrepancies, uneven rims or lateral run-out are made automatically. No manual run-out adjustments are needed.

● **MEASURES ALL ALIGNMENT ANGLES**...Camber, caster, toe, steering-axis inclination, set-back and center-line steering angles are quickly measured. Readings are displayed on large meters for easy observation during alignment. Rear-wheel toe, camber and track may also be measured. Turning-angle readings are made with turning-angle gages.

● **CAMBER & INDIVIDUAL TOE MEA-SURED AUTOMATICALLY**...Motor-driven electro-optical scanner-sensors determine individual toe-settings. Vertical sensing transducers determine camber settings. Settings are recorded on large, clear-view meters in cabinet.

● **MAKES SET-BACK MEASURE-MENTS**...Retro-reflectors positioned on rear wheels reflect beams from front sensors. Enables system to determine front-wheel set-back, rear-wheel track and permits accurate center-line steering setting.

● Use Hunter Electron-A-Line With Any Type of Vehicle Placement Equipment...Floor, Stands, Floor-Rack, Power-Rack, Lift-Rack or Pit-Rack.

Figure 7-16. Typical functions of a modern electronic wheel alignment system. *(Courtesy of Hunter Engineering Co.)*

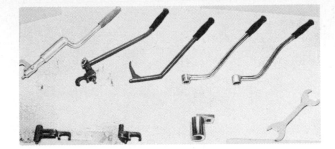

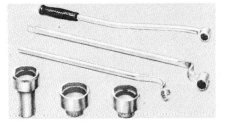

Figure 7-17. A number of special tools are required for wheel alignment, such as those illustrated here. *(Courtesy of Hunter Engineering Co.)*

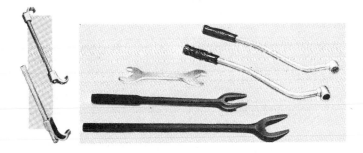

Figure 7-18. Tools such as these are used for adjusting tie rod sleeves. A pipe wrench should not be used. *(Courtesy of Hunter Engineering Co.)*

Figure 7-19. Wheel and tire run-out indicator measures wheel and tire radial and lateral run-out. *(Courtesy of Hunter Engineering Co.)*

Figure 7-20. Steering wheel holder keeps steering wheel centered for adjusting toe. *(Courtesy of Hunter Engineering Co.)*

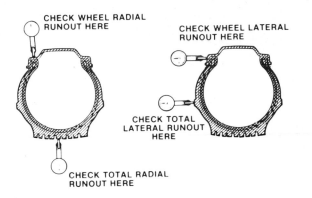

Figure 7-21. Points at which run-out indicator should be positioned for checking radial and lateral run-out of both tire and wheel. *(Courtesy of Ford Motor Co. of Canada Ltd.)*

Figure 7-22. Brake pedal depressor is used to keep brakes applied while checking and adjusting caster. *(Courtesy of Hunter Engineering Co.)*

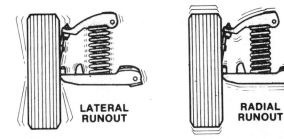

Figure 7-23. Effects of lateral and radial run-out are transmitted to suspension parts and the entire vehicle. *(Courtesy of Ford Motor Co. of Canada Ltd.)*

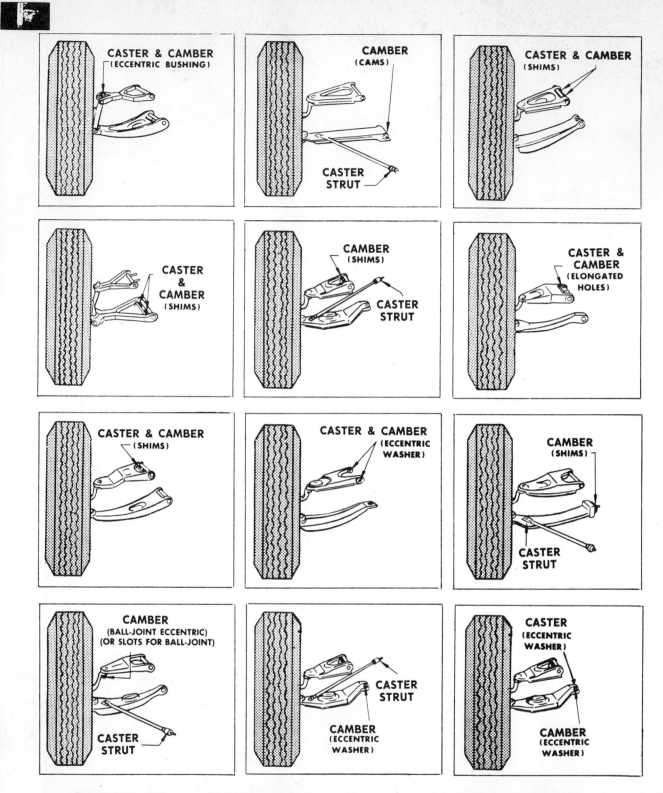

Figure 7-24. Different suspension designs showing various locations for adjusting caster and camber. *(Courtesy of Snap-on Tools Corporation)*

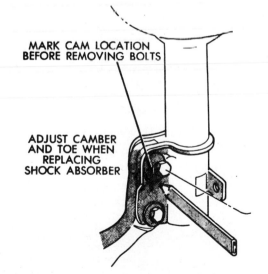

NOTE: TO INCREASE CAMBER, DISCONNECT UPPER BALL JOINT, ROTATE 180° TO POSITION "FLAT" OF FLANGE INBOARD, THEN RECONNECT BALLJOINT.

Figure 7-25. Camber adjustment is provided at the upper ball joint on some vehicles. *(Courtesy of General Motors Corporation)*

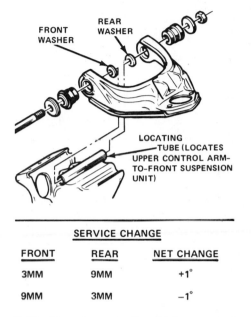

FRONT WASHER

REAR WASHER

LOCATING TUBE (LOCATES UPPER CONTROL ARM-TO-FRONT SUSPENSION UNIT)

SERVICE CHANGE		
FRONT	REAR	NET CHANGE
3MM	9MM	+1°
9MM	3MM	−1°

Figure 7-26. Changing selective thickness washers on upper control arm shaft changes caster as used on some vehicles. *(Courtesy of General Motors Corporation)*

MARK CAM LOCATION BEFORE REMOVING BOLTS

ADJUST CAMBER AND TOE WHEN REPLACING SHOCK ABSORBER

Figure 7-27. Some strut-suspension systems use a cam adjustment for setting camber as shown here. *(Courtesy of Chrylser Corporation)*

Some vehicles equipped with McPherson struts have no means for adjusting caster and camber and if alignment specs are wrong, tires wear excessively. Installation of the KF-39 Caster-Camber Adjusting Kit is made only once and alignment is a very simple adjustment — from under the hood — any time thereafter.

The strut is loosened so a plate can be bolted thru the existing holes in the inner fender. The strut is returned to its original position and held in place by the two plates. The vehicle is now ready for alignment.

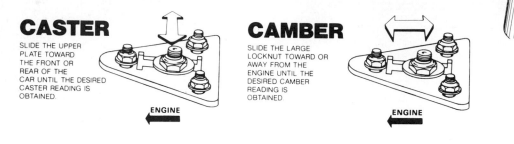

CASTER

SLIDE THE UPPER PLATE TOWARD THE FRONT OR REAR OF THE CAR UNTIL THE DESIRED CASTER READING IS OBTAINED.

← ENGINE

CAMBER

SLIDE THE LARGE LOCKNUT TOWARD OR AWAY FROM THE ENGINE UNTIL THE DESIRED CAMBER READING IS OBTAINED.

← ENGINE

Figure 7-28. *(Courtesy of Moog Automotive Inc.)*

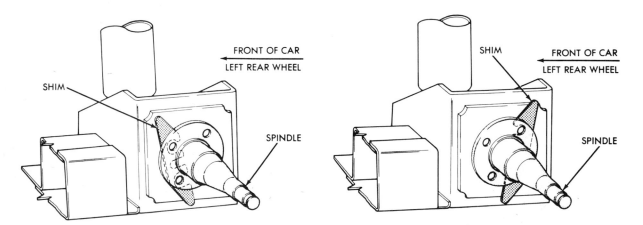

SHIM

FRONT OF CAR
LEFT REAR WHEEL

SPINDLE

SHIM

FRONT OF CAR
LEFT REAR WHEEL

SPINDLE

Figure 7-29. Rear-wheel toe adjustment by means of shims as used on some front-wheel-drive vehicles. *(Courtesy of Chrysler Corporation)*

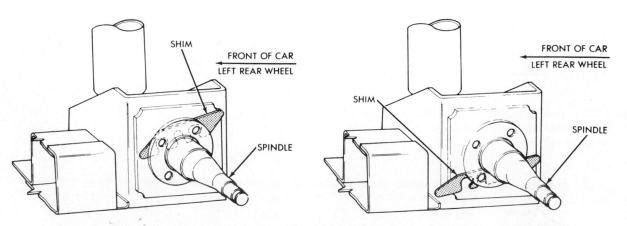

SHIM

FRONT OF CAR
LEFT REAR WHEEL

SPINDLE

FRONT OF CAR
LEFT REAR WHEEL

SHIM

SPINDLE

Figure 7-30. Camber adjustment on rear wheels by means of shims as used on some front-wheel-drive vehicles. *(Courtesy of Chrysler Corporation)*

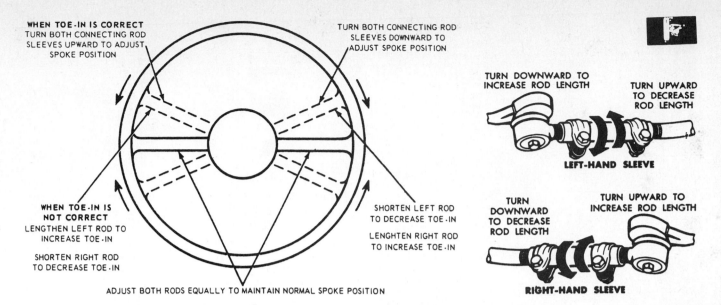

Figure 7-31. How to center steering wheel (left) and adjust steering linkage (toe-in, right). *(Courtesy of Ford Motor Co. of Canada Ltd.)*

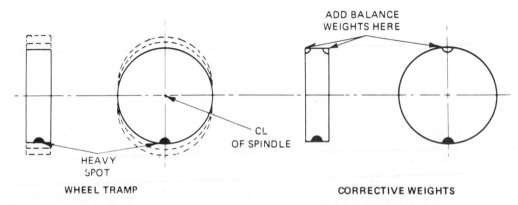

ADD BALANCE
WEIGHTS HERE

CL
OF SPINDLE

HEAVY
SPOT

WHEEL TRAMP

CORRECTIVE WEIGHTS

Figure 7-32. A wheel assembly must be statically in balance to prevent wheel tramp. A wheel that is not statically in balance has a heavy spot. To bring the wheel into static balance, place two equal-sized weights 180° from the heavy spot, one on each side of the wheel rim. The combined weight of the two balance weights must equal the weight of the heavy spot. A statically balanced wheel has equal weight distribution radially around the axis of rotation. *(Courtesy of General Motors Corporation)*

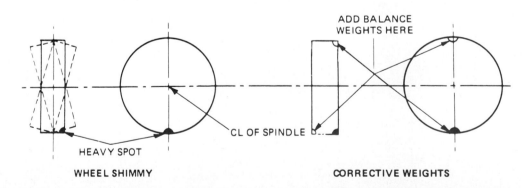

ADD BALANCE
WEIGHTS HERE

CL OF SPINDLE

HEAVY SPOT

WHEEL SHIMMY

CORRECTIVE WEIGHTS

Figure 7-33. A wheel must also be in dynamic balance to prevent wheel shimmy. A wheel that is not in dynamic balance has a heavy spot to one side of the center of the plane of tire rotation. To bring the wheel into dynamic balance, the combined weight of the two balance weights placed as above must equal the heavy spot. A dynamically balanced wheel has equal weight on each side of the wheel center line or plane of rotation. *(Courtesy of General Motors Corporation)*

SPIRIT LEVEL

WEIGHT BEING ADDED

Figure 7-34. Static balancer. This type of balancer does not balance a wheel dynamically. Placing an equal amount of weight on each side of the rim opposite to and equal to the heavy spot will leave dynamic balance unaffected.

Figure 7-35. Strobe-light type of on-a-car balancer balances entire wheel and brake disc or drum assembly as a unit. *(Courtesy of Hunter Engineering Co.)*

Figure 7-36. Electronic computerized wheel balancer balances wheels both statically and dynamically. *(Courtesy of Sun Electric Corporation)*

PROJECT NO. 7-A

**TITLE: <u>WHEEL ALIGNMENT
AND BALANCE</u>**

STUDENT'S NAME _____

PERFORMANCE OBJECTIVES

After sufficient opportunity to study this portion of the text and the appropriate training models and with the instructor's supervision and demonstrations, you should be able to perform the following tasks at the request of your instructor.

TASK 1---State the purpose of wheel alignment. Name the five alignment angles and define each angle. Describe the effects of correct and incorrect alignment angles.

TASK 2---State the purpose of wheel balance. Define static balance and dynamic balance. Describe the effects of static unbalance and dynamic unbalance.

PERFORMANCE EVALUATION

Your instructor may require you to perform these tasks in any of the following ways in order to evaluate your performance:
- By asking test questions
- By asking you to describe the performance of these tasks in writing
- By asking you to describe the performance of these tasks orally

TITLE: <u>WHEEL ALIGNMENT</u>
<u>AND BALANCE</u>
<u>DIAGNOSIS AND</u>
<u>PROCEDURE</u>

STUDENT'S NAME _____

PERFORMANCE OBJECTIVES

After sufficient opportunity to study this portion of the text and the appropriate training models and with the instructor's supervision and demonstrations, you should be able to perform the following tasks at the request of your instructor.

TASK 1--- Complete the wheel alignment and balance diagnostic chart provided. Diagnose wheel alignment and balance problems on a vehicle specified by your instructor.

TASK 2--- Perform a complete prealignment inspection and a complete wheel alignment on a vehicle specified by your instructor.

TASK 3--- Balance the wheels on a vehicle specified by your instructor.

PERFORMANCE EVALUATION

Your ability to perform these tasks will be evaluated by your instructor on the basis of how accurately you have followed the procedures and specifications given to you by your instructor, by the text, and by the manufacturer's service manual.

PROBLEM	CAUSE	CORRECTION
Hard steering	1. 2. 3. 4. 5.	1. 2. 3. 4. 5.
Car pulls to one side	1. 2. 3. 4. 5.	1. 2. 3. 4. 5.
Too much play in steering	1. 2. 3. 4.	1. 2. 3. 4.
Front-wheel shimmy	1. 2. 3. 4. 5.	1. 2. 3. 4. 5.
Front-end noise	1. 2. 3. 4. 5.	1. 2. 3. 4. 5.
Instability	1. 2. 3. 4. 5.	1. 2. 3. 4. 5.

PREALIGNMENT INSPECTION

— TIRES   OK _____   NOT OK _____   EXPLAIN _____

— TIRE PRESSURES   OK _____   NOT OK _____   EXPLAIN _____

— CURB HEIGHT   OK _____   NOT OK _____   EXPLAIN _____

— TRACKING   OK _____   NOT OK _____   EXPLAIN _____

— WHEEL BEARINGS   OK _____   NOT OK _____   EXPLAIN _____

— TIRE RUN-OUT   L.F. _____   R.F. _____   L.R. _____   R.R. _____

— WHEEL RUN-OUT   L.F. _____   R.F. _____   L.R. _____   R.R. _____

— STEERING LINKAGE   OK _____   NOT OK _____   EXPLAIN _____

— FRONT SUSPENSION   OK _____   NOT OK _____   EXPLAIN _____

— REAR SUSPENSION   OK _____   NOT OK _____   EXPLAIN _____

— STEERING GEAR   OK _____   NOT OK _____   EXPLAIN _____

— LUBE LEVEL   OK _____   NOT OK _____

— POWER-STEERING BELT   OK _____   NOT OK _____

— LEAKAGE   OK _____   NOT OK _____

— POWER-STEERING ASSIST   OK _____   NOT OK _____   EXPLAIN _____

PREALIGNMENT ANGLE CHECKS

— CAMBER   LEFT _____   RIGHT _____

— CASTER   LEFT _____   RIGHT _____

— SAI   LEFT _____   RIGHT _____

— TOE-OUT ON TURNS _____

CORRECTED ALIGNMENT ANGLES

— CAMBER

   LEFT   ACTUAL_____   SPECIFIED _____

   RIGHT   ACTUAL_____   SPECIFIED _____

-- CASTER

   LEFT   ACTUAL_____   SPECIFIED _____

   RIGHT   ACTUAL_____   SPECIFIED _____

Steering Axis Inclination

Steering axis inclination is the inward tilt of the steering knuckle at the top. Steering axis inclination is measured in degrees and is not adjustable. If incorrect, suspension parts are at fault and must be replaced. Steering axis inclination is needed for the following purposes:

1. To place the weight of the vehicle more nearly under the center of the tire where it contacts the road

2. To reduce steering effort, because of item 1

3. To provide good directional control (Figure 7-10)

Incorrect steering axis inclination will affect all the preceding and may prevent being able to set camber properly.

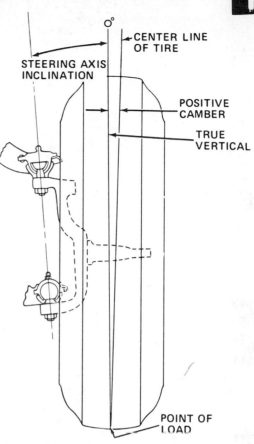

Figure 7-8. Relationship of steering axis inclination to camber. One of the effects of these two angles is to place the point of vehicle load at the center of the tire on the road surface. *(Courtesy of General Motors Corporation)*

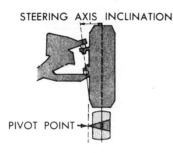

Figure 7-7. Steering axis inclination is the inward tilt of the spindle support at the top. *(Courtesy of Chrysler Corporation)*

Toe-In

Toe-in occurs when the front wheels are slightly closer together at the front than at the rear. Toe-in is measured in inches, millimeters, or degrees. A limited amount of toe-in or toe-out is needed to allow for the fact that the wheels spread apart or come

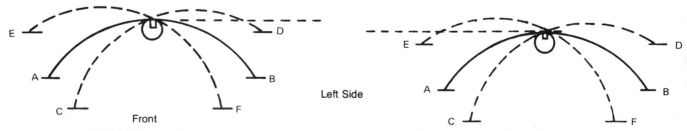

Figure 7-9. The effect of steering axis inclination on the travel of the outer end of spindle as shown by the line A to B when front wheels are turned from right to left. Since the spindle cannot in fact push the wheel into the road surface, the result is that the car is raised when turned to right or left. This helps return front wheels to straight head after a turn and aids directional stability. The line A to B represents a zero caster setting, the line C to D a negative caster setting, and the line E to F a positive caster setting. A car will lead to the side with the most negative or least positive caster. The figure on the left represents a zero camber setting; the figure on the right shows the result of a positive camber setting. The effect is that it tries to lower the outer end of the spindle.

**KING PIN INCLINATION
ANGLE HELPS WHEELS TO RETURN
WHEN COMING OUT OF TURNS.**

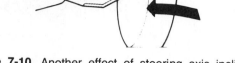

Figure 7-10. Another effect of steering axis inclination is shown here. *(Courtesy of Chrysler Corporation)*

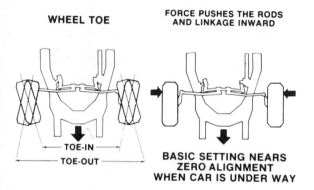

WHEEL TOE

**FORCE PUSHES THE RODS
AND LINKAGE INWARD**

TOE-IN
TOE-OUT

**BASIC SETTING NEARS
ZERO ALIGNMENT
WHEN CAR IS UNDER WAY**

Figure 7-11. The objective in setting toe is to have the front wheels running straight ahead (zero running toe). To achieve this, a small amount of static toe-in must be present. Driving force deflects steering parts enough to result in zero running toe. Some cars with a negative scrub radius may require a toe-out setting. *(Courtesy of Ford Motor Co. of Canada Ltd.)*

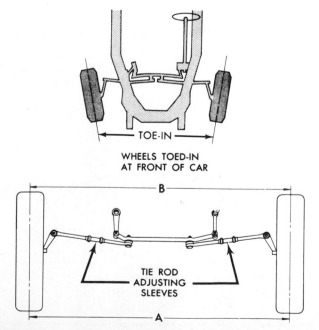

TOE-IN

**WHEELS TOED-IN
AT FRONT OF CAR**

B

TIE ROD
ADJUSTING
SLEEVES

A

Figure 7-12. When front wheels have toe-in, they are closer together at the front wheels than at the rear. Dimension A is less than dimension B. *(Courtesy of General Motors Corporation and Chrysler Corporation)*

together slightly at the front when driving down the road, depending on vehicle design. This provides a zero running toe and no tire scuffing.

Incorrect toe-in or toe-out is the most frequent cause of rapid tire-tread wear. Toe setting is the last adjustment to be made when performing a wheel alignment. On most front-wheel-drive vehicles, toe-out setting is required to provide a zero running toe. This is because the driving front wheels are trying to go around the SAI pivot point with a negative scrub radius.

Toe-Out on Turns

Toe-out on turns is the different turning radius of the two front wheels. When the car is in a turn, the inner wheel is turned more than the outer wheel, resulting in toe-out on turns. This is caused by the steering arms being bent inward where they connect to the steering linkage. It is needed to prevent tire scuffing (dragging sideways) during a turn. Since the inner wheel follows a smaller circle than the outer wheel when in a turn, toe-out on turns is necessary. Toe-out on turns is not adjustable and is corrected by replacing steering arms.

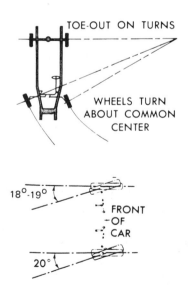

TOE-OUT ON TURNS

WHEELS TURN
ABOUT COMMON
CENTER

18°-19°

FRONT
OF
CAR

20°

Figure 7-13. Toe-out on turns (turning radius) is needed to reduce tire scuffing when turning corner. Since all four wheels turn around a common center, the inner front wheel must be turned sharper, as shown above. This is accomplished by steering arm design. Steering arms are angled inward where they attach to the steering linkage. *(Courtesy of General Motors Corporation and Chrysler Corporation)*

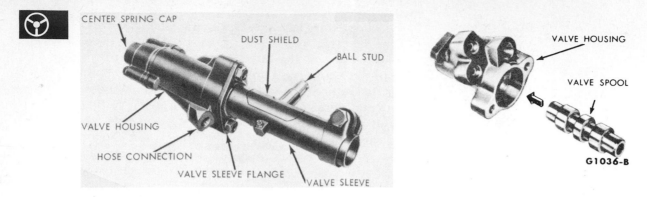

Figure 6-24. Major components of linkage booster power-steering control valve assembly. *(Courtesy of Ford Motor Co. of Canada Ltd.)*

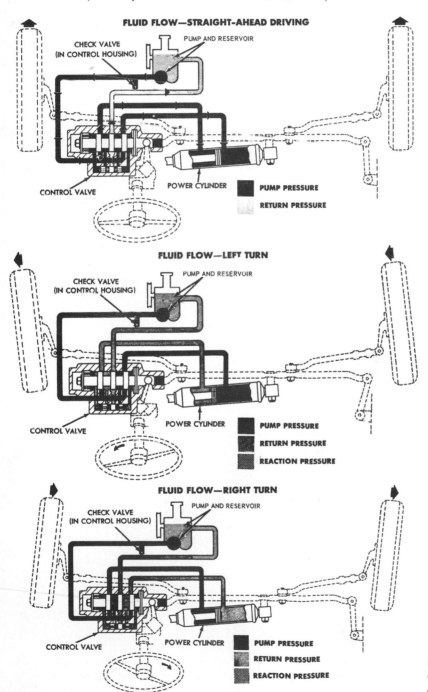

Figure 6-25. Operation of linkage booster type of power steering. *(Courtesy of Ford Motor Co. of Canada Ltd.)*

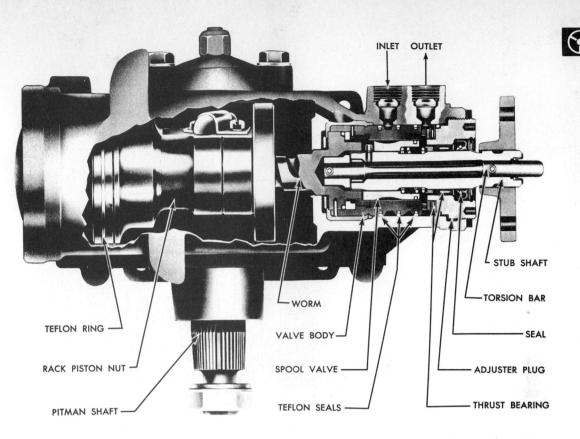

INLET   OUTLET

STUB SHAFT

TORSION BAR

SEAL

ADJUSTER PLUG

THRUST BEARING

WORM

VALVE BODY

SPOOL VALVE

TEFLON SEALS

TEFLON RING

RACK PISTON NUT

PITMAN SHAFT

Figure 6-22. Integral power steering with a torsion bar and different type of control valve assembly. *(Courtesy of General Motors Corporation)*

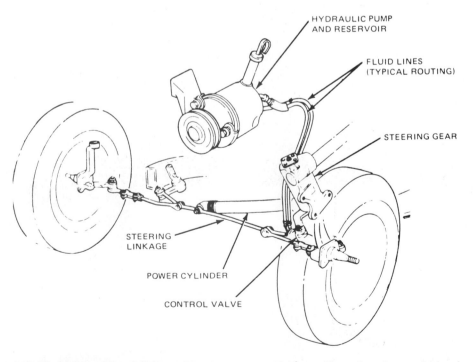

HYDRAULIC PUMP
AND RESERVOIR

FLUID LINES
(TYPICAL ROUTING)

STEERING GEAR

STEERING
LINKAGE

POWER CYLINDER

CONTROL VALVE

Figure 6-23. Major parts of linkage booster power steering. Note that the control valve, power cylinder, and steering gear are separate units. The power cylinder provides steering assist. *(Courtesy of Ford Motor Co. of Canada Ltd.)*

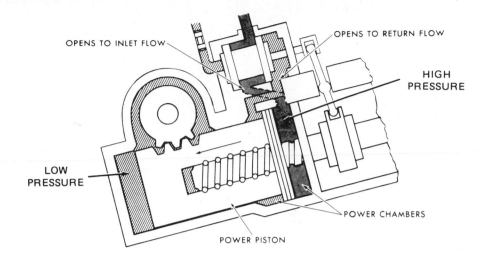

Figure 6-20. During right turn the power piston moves down, creating an opposite reaction on the control valve as compared to Figure 6-19. The higher pressure above the piston helps move the piston down to the low-pressure area, providing steering assist. *(Courtesy of Chrysler Corporation)*

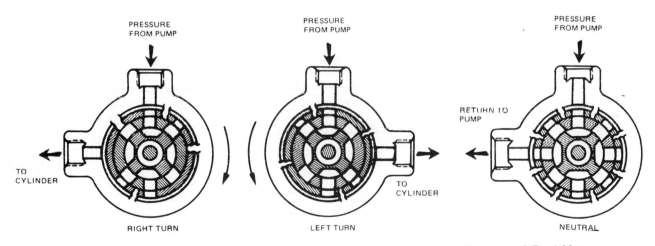

Figure 6-21. Hydraulic pressure flow paths of rotary-type control valve. *(Courtesy of Ford Motor Co. of Canada Ltd.)*

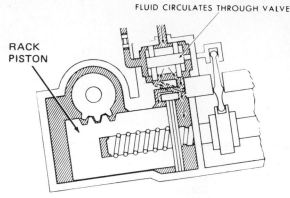

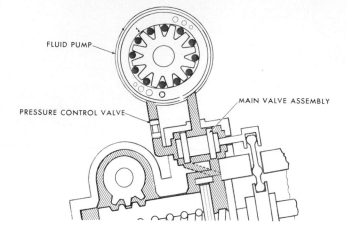

Figure 6-16. The power-steering pump (fluid pump) provides the hydraulic pressure for the power steering gear. Pressure is regulated by the pressure control valve and is directed to the rack piston by the main valve assembly. This provides power-assisted steering whenever the steering wheel is turned. *(Courtesy of Chrysler Corporation)*

Figure 6-17. During straight-ahead driving, hydraulic pressure is directed equally to both sides of the rack piston. *(Courtesy of Chrysler Corporation)*

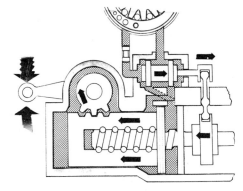

Figure 6-18. If a front wheel hits a bump or hole in the road, the front wheels are reflected to the right or left. This creates a tendency to self-steer to right or left. This action is transferred to the control valve as shown by arrows causing hydraulic pressure to offset the self-steering tendency. *(Courtesy of Chrysler Corporation)*

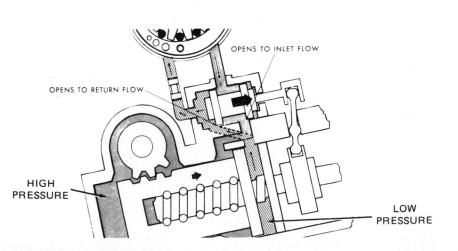

Figure 6-19. A left turn causes the worm shaft to thread into the rack piston, which pulls the piston up. A reaction on the pivot moves the control valve up, which directs hydraulic pressure to the bottom side of the rack piston. At the same time a return passage is opened from the chamber above the piston to allow fluid return and a pressure drop in this chamber. In effect, the pressure difference on the two sides of the piston assists the piston to move up which provides steering assist. *(Courtesy of Chrysler Corporation)*

— SAI

    LEFT   ACTUAL_____   SPECIFIED _____

    RIGHT   ACTUAL_____   SPECIFIED _____

— TOE-OUT ON TURNS   ACTUAL_____   SPECIFIED _____

— TOE-IN   ACTUAL_____   SPECIFIED _____

STUDENT SIGN HERE _____

INSTRUCTOR VERIFY _____

Self-Check

1. What is the purpose of wheel alignment and balance?

2. What six alignment factors must be checked and corrected for proper alignment?

3. List three effects of incorrect camber.

4. Give three reasons why caster is needed.

5. How is incorrect steering axis inclination corrected?

6. What is the result of excessive toe-in or toe-out?

7. List eight pre-alignment checks that should be made before wheel alignment.

8. What is the cause of wheel tramp?

9. What is the cause of wheel shimmy?

10. List three motheds of adjusting camber on strut-suspension systems.

Performance Evaluation

After study of this chapter and sufficient opportunity to practice wheel alignment and balance, you should be able to do the following.

1. Follow the accepted general precautions.

2. Accurately perform all pre-alignment checks.

3. Accurately measure and correct all alignment factors to the manufacturer's specifications.

4. Properly prepare the wheel assemblies for balancing.

5. Accurately balance the wheel assemblies both statically and dynamically.

6. Perform the necessary checks to determine the success of the alignment and balance procedures.

7. Properly prepare the vehicle for customer acceptance.

8. Complete the Self-Check with at least 80 percent accuracy.

9. Complete all practical work with 100 percent accuracy.

Conversion Charts

TAPS · DIES · GAGES · SCREW PLATES · SCREW EXTRACTORS

DECIMAL EQUIVALENTS AND TAP DRILL SIZES

DRILL SIZE	DECIMAL	TAP SIZE	DRILL SIZE	DECIMAL	TAP SIZE	DRILL SIZE	DECIMAL	TAP SIZE
1/64	.0156		17	.1730		Q	.3320	3/8-24
1/32	.0312		16	.1770	12-24	R	.3390	
60	.0400		15	.1800		11/32	.3437	
59	.0410		14	.1820	12-28	S	.3480	
58	.0420		13	.1850	12-32	T	.3580	
57	.0430		3/16	.1875		23/64	.3594	
56	.0465		12	.1890		U	.3680	7/16-14
3/64	.0469	0-80	11	.1910		3/8	.3750	
55	.0520		10	.1935		V	.3770	
54	.0550	1-56	9	.1960		W	.3860	
53	.0595	1-64, 72	8	.1990		25/64	.3906	7/16-20
1/16	.0625		7	.2010	1/4-20	X	.3970	
52	.0635		13/64	.2031		Y	.4040	
51	.0670		6	.2040		13/32	.4062	
50	.0700	2-56, 64	5	.2055		Z	.4130	
49	.0730		4	.2090		27/64	.4219	1/2-13
48	.0760		3	.2130	1/4 28	7/16	.4375	
5/64	.0781		7/32	.2187		29/64	.4531	1/2-20
47	.0785	3-48	2	.2210		15/32	.4687	
46	.0810		1	.2280		31/64	.4844	9/16-12
45	.0820	3-56, 4-32	A	.2340		1/2	.5000	
44	.0860	4-36	15/64	.2344		33/64	.5156	9/16-18
43	.0890	4-40	B	.2380		17/32	.5312	5/8-11
42	.0935	4-48	C	.2420		35/64	.5469	
3/32	.0937		D	.2460		9/16	.5625	
41	.0960		E, 1/4	.2500		37/64	.5781	5/8-18
40	.0980		F	.2570	5/16-18	19/32	.5937	11/16-11
39	.0995		G	.2610		39/64	.6094	
38	.1015	5-40	17/64	.2656		5/8	.6250	11/16-16
37	.1040	5-44	H	.2660		41/64	.6406	
36	.1065	6-32	I	.2720	5/16-24	21/32	.6562	3/4-10
7/64	.1093		J	.2770		43/64	.6719	
35	.1100		K	.2810		11/16	.6875	3/4-16
34	.1110	6-36	9/32	.2812		45/64	.7031	
33	.1130	6-40	L	.2900		23/32	.7187	
32	.1160		M	.2950		47/64	.7344	
31	.1200		19/64	.2968		3/4	.7500	
1/8	.1250		N	.3020		49/64	.7656	7/8-9
30	.1285		5/16	.3125	3/8-16	25/32	.7812	
29	.1360	8-32, 36	O	.3160		51/64	.7969	
28	.1405	8-40	P	.3230		13/16	.8125	7/8-14
9/64	.1406		21/64	.3281		53/64	.8281	
27	.1440					27/32	.8437	
26	.1470					55/64	.8594	
25	.1495	10-24				7/8	.8750	1-8
24	.1520					57/64	.8906	
23	.1540					29/32	.9062	
5/32	.1562					59/64	.9219	
22	.1570	10-30				15/16	.9375	1-12,14
21	.1590	10-32				61/64	.9531	
20	.1610					31/32	.9687	
19	.1660					63/64	.9844	
18	.1695					1	1.000	
11/64	.1719							

PIPE THREAD SIZES

THREAD	DRILL	THREAD	DRILL
1/8-27	R	1 1/2-11 1/2	1 47/64
1/4-18	7/16	2-11 1/2	2 7/32
3/8-18	37/64	2 1/2-8	2 5/8
1/2-14	23/32	3-8	3 1/4
3/4-14	59/64	3 1/2-8	3 3/4
1-11 1/2	1 5/32	4-8	4 1/4
1 1/4-11 1/2	1 1/2		

DIMENSION AND TEMPERATURE CONVERSION CHART

Inches	Decimals	Milli-meters	Inches to millimeters		Millimeters to inches		Fahrenheit & Celsius			
			Inches	mm	mm	Inches	°F	°C	°C	°F
1/64	.015625	.3969	.0001	.00254	0.001	.000039	-20	-28.9	-30	-22
1/32	.03125	.7937	.0002	.00508	0.002	.000079	-15	-26.1	-28	-18.4
3/64	.046875	1.1906	.0003	.00762	0.003	.000118	-10	-23.3	-26	-14.8
1/16	.0625	1.5875	.0004	.01016	0.004	.000157	-5	-20.6	-24	-11.2
5/64	.078125	1.9844	.0005	.01270	0.005	.000197	0	-17.8	-22	-7.6
3/32	.09375	2.3812	.0006	.01524	0.006	.000236	1	-17.2	-20	-4
7/64	.109375	2.7781	.0007	.01778	0.007	.000276	2	-16.7	-18	-0.4
1/8	.125	3.1750	.0008	.02032	0.008	.000315	3	-16.1	-16	3.2
9/64	.140625	3.5719	.0009	.02286	0.009	.000354	4	-15.6	-14	6.8
5/32	.15625	3.9687	.001	.0254	0.01	.00039	5	-15.0	-12	10.4
11/64	.171875	4.3656	.002	.0508	0.02	.00079	10	-12.2	-10	14
3/16	.1875	4.7625	.003	.0762	0.03	.00118	15	-9.4	-8	17.6
13/64	.203125	5.1594	.004	.1016	0.04	.00157	20	-6.7	-6	21.2
7/32	.21875	5.5562	.005	.1270	0.05	.00197	25	-3.9	-4	24.8
15/64	.234375	5.9531	.006	.1524	0.06	.00236	30	-1.1	-2	28.4
1/4	.25	6.3500	.007	.1778	0.07	.00276	35	1.7	0	32
17/64	.265625	6.7469	.008	.2032	0.08	.00315	40	4.4	2	35.6
9/32	.28125	7.1437	.009	.2286	0.09	.00354	45	7.2	4	39.2
19/64	.296875	7.5406	.01	.254	0.1	.00394	50	10.0	6	42.8
5/16	.3125	7.9375	.02	.508	0.2	.00787	55	12.8	8	46.4
21/64	.328125	8.3344	.03	.762	0.3	.01181	60	15.6	10	50
11/32	.34375	8.7312	.04	1.016	0.4	.01575	65	18.3	12	53.6
23/64	.359375	9.1281	.05	1.270	0.5	.01969	70	21.1	14	57.2
3/8	.375	9.5250	.06	1.524	0.6	.02362	75	23.9	16	60.8
25/64	.390625	9.9219	.07	1.778	0.7	.02756	80	26.7	18	64.4
13/32	.40625	10.3187	.08	2.032	0.8	.03150	85	29.4	20	68
27/64	.421875	10.7156	.09	2.286	0.9	.03543	90	32.2	22	71.6
7/16	.4375	11.1125	.1	2.54	1	.03937	95	35.0	24	75.2
29/64	.453125	11.5094	.2	5.08	2	.07874	100	37.8	26	78.8
15/32	.46875	11.9062	.3	7.62	3	.11811	105	40.6	28	82.4
31/64	.484375	12.3031	.4	10.16	4	.15748	110	43.3	30	86
1/2	.5	12.7000	.5	12.70	5	.19685	115	46.1	32	89.6
33/64	.515625	13.0969	.6	15.24	6	.23622	120	48.9	34	93.2
17/32	.53125	13.4937	.7	17.78	7	.27559	125	51.7	36	96.8
35/64	.546875	13.8906	.8	20.32	8	.31496	130	54.4	38	100.4
9/16	.5625	14.2875	.9	22.86	9	.35433	135	57.2	40	104
37/64	.578125	14.6844	1	25.4	10	.39370	140	60.0	42	107.6
19/32	.59375	15.0812	2	50.8	11	.43307	145	62.8	44	112.2
39/64	.609375	15.4781	3	76.2	12	.47244	150	65.6	46	114.8
5/8	.625	15.8750	4	101.6	13	.51181	155	68.3	48	118.4
41/64	.640625	16.2719	5	127.0	14	.55118	160	71.1	50	122
21/32	.65625	16.6687	6	152.4	15	.59055	165	73.9	52	125.6
43/64	.671875	17.0656	7	177.8	16	.62992	170	76.7	54	129.2
11/16	.6875	17.4625	8	203.2	17	.66929	175	79.4	56	132.8
45/64	.703125	17.8594	9	228.6	18	.70866	180	82.2	58	136.4
23/32	.71875	18.2562	10	254.0	19	.74803	185	85.0	60	140
47/64	.734375	18.6531	11	279.4	20	.78740	190	87.8	62	143.6
3/4	.75	19.0500	12	304.8	21	.82677	195	90.6	64	147.2
49/64	.765625	19.4469	13	330.2	22	.86614	200	93.3	66	150.8
25/32	.78125	19.8437	14	355.6	23	.90551	205	96.1	68	154.4
51/64	.796875	20.2406	15	381.0	24	.94488	210	98.9	70	158
13/16	.8125	20.6375	16	406.4	25	.98425	212	100.0	75	167
53/64	.828125	21.0344	17	431.8	26	1.02362	215	101.7	80	176
27/32	.84375	21.4312	18	457.2	27	1.06299	220	104.4	85	185
55/64	.859375	21.8281	19	482.6	28	1.10236	225	107.2	90	194
7/8	.875	22.2250	20	508.0	29	1.14173	230	110.0	95	203
57/64	.890625	22.6219	21	533.4	30	1.18110	235	112.8	100	212
29/32	.90625	23.0187	22	558.8	31	1.22047	240	115.6	105	221
59/64	.921875	23.4156	23	584.2	32	1.25984	245	118.3	110	230
15/16	.9375	23.8125	24	609.6	33	1.29921	250	121.1	115	239
61/64	.953125	24.2094	25	635.0	34	1.33858	255	123.9	120	248
31/32	.96875	24.6062	26	660.4	35	1.37795	260	126.6	125	257
63/64	.984375	25.0031	27	690.6	36	1.41732	265	129.4	130	266

METRIC—ENGLISH CONVERSION TABLE

Multiply	by	to get equivalent number of:
LENGTH		
Inch	25.4	millimetres (mm)
Foot	0.304 8	metres (m)
Yard	0.914 4	metres
Mile	1.609	kilometres (km)
AREA		
Inch^2	645.2	millimetres^2 (mm^2)
	6.45	centimetres^2 (cm^2)
Foot^2	0.092 9	metres^2 (m^2)
Yard^2	0.836 1	metres^2
VOLUME		
Inch^3	16 387.	mm^3
	16.387	cm^3
	0.016 4	litres (l)
Quart	0.946 4	litres
Gallon	3.785 4	litres
Yard^3	0.764 6	metres^3 (m^3)
MASS		
Pound	0.453 6	kilograms (kg)
Ton	907.18	kilograms (kg)
Ton	0.907	tonne (t)
FORCE		
Kilogram	9.807	newtons (N)
Ounce	0.278 0	newtons
Pound	4.448	newtons
TEMPERATURE		
Degree Fahrenheit	($^{\circ}$F-32)÷ 1.8	degree Celsius

Multiply	by	to get equivalent number of:
ACCELERATION		
Foot/sec^2	0.304 8	metre/sec^2 (m/s^2)
Inch/sec^2	0.025 4	metre/sec^2
TORQUE		
Pound-inch	0.112 98	newton-metres (N-m)
Pound-foot	1.355 8	newton-metres
POWER		
Horsepower	0.746	kilowatts (kW)
PRESSURE OR STRESS		
Inches of water	0.249 1	kilopascals (kPa)
Pounds/sq. in.	6.895	kilopascals
ENERGY OR WORK		
BTU	1 055.	joules (J)
Foot-pound	1.355 8	joules
Kilowatt-hour	3 600 000. or 3.6 x 10^6	joules (J = one W's)
LIGHT		
Foot candle	1.076 4	lumens/metre^2 (lm/m^2)
FUEL PERFORMANCE		
Miles/gal	0.425 1	kilometres/litre (km/l)
Gal/mile	2.352 7	litres/kilometre (l/km)
VELOCITY		
Miles/hour	1.609 3	kilometres/hr. (km/h)

CONVERSION-ENGLISH AND SI METRIC MEASURE

Cubic Centimeters to Inches:

When changing cubic centimeters to cubic inches, multiply cubic centimeters times .061 to obtain cubic inches, (C.C. × .061 = Cubic Inches).

Cubic Inches to Centimeters:

When changing cubic inches to cubic centimeters, multiply cubic inches times 16.93 to obtain cubic centimeters, (Cubic Inches × 16.39 = C.C.).

Liters to Cubic Inches:

When changing liters to cubic inches, multiply liters times 61.02 to obtain cubic inches, (Liters × 61.02 = Cubic Inches).

Cubic Inches to Liters:

When changing cubic inches to liters, multiply cubic inches times .01639 to obtain liters, (Cubic Inches × .01639 = Liters).

Cubic Centimeters to Liters

When changing cubic centimetres to liters, divide by 1,000 simply by moving the decimal point three figures to the left.

Liters to Cubic Centimeters:

When changing liters to cubic centimeters, move the decimal point three figures to the right.

Miles to Kilometers:

When changing miles to kilometers, multiply miles times 1.609 to obtain kilometers, (Miles × 1.609 = Kilometers).

Kilometers to Miles:

When changing kilometers to miles, multiply kilometers times .6214 to obtain miles, (Kilometers × .6214 = Miles).

Pounds to Kilograms:

When changing pounds to kilograms, multiply pounds times .4536 to obtain kilograms, (Pounds × .4536 = Kilograms).

Kilograms to Pounds:

When changing kilograms to pounds, multiply kilograms times 2.2046 to obtain pounds, (Kilograms × 2.2046 = Pounds).

Pounds to Newtons:

When changing pounds to newtons, multiply pounds times 4.4482 to obtain newtons, (Pounds × 4.4482 = Newtons).

Newtons to Pounds:

When changing newtons to pounds, multiply newtons times .2248 to obtain pounds, (Newtons × .2248 = Pounds).

Pound-Feet to Newton Meters:

When changing pound-feet to newton meters, multiply pound-feet times 1.3558 to newton meters, (Pound-Feet × 1.3558 = Newton Meters).

Newton Meters to Pound-Feet:

When changing newton meters to pound-feet, multiply newton meters times .7376 to obtain pounds-feet, (Newton Meters × .7376 = Pound-Feet).

Pounds Per Square Inch to Kilopascals:

When changing pounds per square inch to kilopascals, multiply pounds per square inch times 6.895 to obtain kilopascals, (pounds Per Square Inch × 6.895 = Kilopascals.).

Kilopascals to Pounds Per Square Inch:

When changing kilopascals to pounds per square inch, multiply kilopascals times .1450 to obtain pounds per square inch, (Kilopascals × .1450 = Pounds Per Square Inch).

Index